Philosophies of Technologies

Innovation and Responsibility Set

coordinated by
Bernard Reber and Robert Gianni

Volume 11

Philosophies of Technologies

Theory as Practice

Edited by

Valérie Charolles
Élise Lamy-Rested

iSTE

WILEY

First published 2024 in Great Britain and the United States by ISTE Ltd and John Wiley & Sons, Inc.

ISTE Ltd
27-37 St George's Road
London SW19 4EU
UK

www.iste.co.uk

John Wiley & Sons, Inc.
111 River Street
Hoboken, NJ 07030
USA

www.wiley.com

Library of Congress Control Number: 2023947688

British Library Cataloguing-in-Publication Data
A CIP record for this book is available from the British Library
ISBN 978-1-78630-870-2

Contents

Author Presentation

Bruno Bachimont – Professor at the Université de Technologie de Compiègne (Compiègne University of Technology), attached to the CosTech research unit. He is interested in the changes brought about by digital technology in knowledge and approaches this matter from a knowledge engineering (ontologies and Web of data), documentary engineering (archives and heritage) and digital philosophy (epistemology, memory and technology) perspective. He has been a Research Director and Scientific Advisor at the Institut national de l'audiovisuel (INA) and a Research Director at UTC and Sorbonne University (Faculty of Science and Engineering).

Pierre Caye – As Research Director at the CNRS, he devotes his research to artistic theories from the Renaissance to the Enlightenment, with a particular focus on architecture. In 2012, he founded the Groupement de recherche international sur "les savoirs artistiques et les traités d'art de la Renaissance aux Lumières" (International Research Group on "Artistic Knowledge and Art Treatises from the Renaissance to the Enlightenment"). In collaboration with Françoise Choay, he translated and commented on Alberti's *L'Art d'édifier* (Le Seuil, 2004). More recently, he has also published two works on sustainable development: *Critique de la destruction créatrice. Production et humanisme* (Les Belles Lettres, 2015) and *Durer. Éléments pour la transformation du système productif* (Les Belles Lettres, 2020).

Pierre-Antoine Chardel – Senior researcher at the Laboratoire d'anthropologie politique (UMR 8177, CNRS/EHESS) and professor of social sciences and ethics at Institut Mines-Télécom Business School (IMT-BS). His work focuses on the ethical and socio-philosophical issues of the hypermodern era. He has recently been a visiting researcher at the MediaLab of Sciences Po Paris. His recent publications include *L'Empire du signal. De l'écrit aux écrans* (CNRS Éditions, 2020) and *Socio-philosophie des technologies numériques. Éthique, société, organisations* (Presses des Mines, 2022).

Valérie Charolles – Senior researcher at the Laboratoire d'anthropologie politique (UMR 8177, CNRS/EHESS); her work focuses on the role of economy, quantification and technology in the contemporary constitution of the subject and the world. She holds a PhD in philosophy and is a qualified lead researcher (HDR). She has published five books, including *Philosophie de l'écran* (Fayard, 2013) and *Se libérer de la domination des chiffres* (Fayard, 2022).

Andrew Feenberg – Philosopher and past holder of the Canada Research Chair in Philosophy of Technology at Simon Fraser University. He was Program Director at the Collège international de philosophie from 2013 to 2019. His research focuses on the philosophy of technology, the Internet and the Frankfurt School. He has written numerous articles and books on the philosophy of technology. His recent books include *The Ruthless Critique of Everything Existing: Nature and Revolution in Marcuse's Philosophy of Praxis*, (Verso, 2023) and *Technosystem: The Social Life of Reason* (Harvard University Press, 2017)[1].

Xavier Guchet – Professor at the Université de Technologie de Compiègne, Director of the Knowledge, Organizations and Technical Systems Laboratory (CosTech UR 2223). His work focuses on the philosophy of technology as a specific field of philosophy, as well as on the epistemological, ethical and political issues of recent technological developments: nanotechnologies, personalized medicine and tissue engineering. He has published five books, including *Care in Technology* (ISTE Ltd and Wiley, 2021).

1 For more information, visit his website: www.sfu.ca/~andrewf.

Éric Guichard – At the Triangle Laboratory (CNRS/ENS Lyon) and IXXI (Institut des systèmes complexes de Rhône-Alpes), Éric Guichard is developing a digital philosophy based on writing and materiality. He teaches at ENSSIB and was Program Director at the Collège international de philosophie[2].

Élise Lamy-Rested – Holds a PhD in philosophy from Paris-Sorbonne University and is a Marie Sklodowska-Curie researcher (MSC Cofund H2020, SASPRO2) at the Institute of Philosophy of the Slovak Academy of Sciences, following a first Marie Sklodowska-Curie postdoctorate (MSC Cofund H2020) at the Université libre de Bruxelles (2019–2022). She was also formerly Program Director at the Collège international de philosophie (2016–2022). Her current research focuses on "technical life". She has published two books, including *Derrida. Excès de vie* (Kimé, 2017), as well as numerous articles.

Jean Lassègue – Philosopher, senior research scientist at the CNRS and a member of the Centre Georg Simmel – Recherches franco-allemandes en sciences sociales (EHESS – CNRS UMR 8131). Connecting epistemology, anthropology and history, his research focuses on symbolic mediations, particularly language and number writing systems that enable the collective elaboration of knowledge, from the natural to the legal sciences.

Anna Longo – Holds a doctorate in aesthetics and is a program director at the Collège international de philosophie. Her research focuses on the history and limits of the current system of automated information production. She is the author of *Le jeu de l'induction. Automatisation de la connaissance et réflexion philosophique* (Mimesis, 2022).

Vincent Puig – Director of the Institut de recherche et d'innovation at the Centre Pompidou, which he founded in 2006 with philosopher Bernard Stiegler. As a digital theorist and practitioner, he has led a number of research projects at IRCAM, the Centre Pompidou and the IRI, notably on indexing, annotation, categorization and editorialization technologies. Today, he coordinates contributory research projects, mainly in Seine-Saint-Denis, with a focus on the contribution economy.

2 His work is available at the following website: http://barthes.enssib.fr.

Bernard Reber – Research Director at the CNRS, as well as a moral and political philosopher. He is a member of the Centre de recherches politiques de Sciences Po (Paris) and the Centre des politiques de la terre (Paris Cité University). His work combines moral and political philosophy, as well as sociology, applied ethics and political theory, to address concepts such as deliberation, argumentation, pluralism, responsibility and the precautionary principle. For over 20 years, he has analyzed experiments in participatory technology assessment and, more recently, citizens' climate assemblies.

Tyler Reigeluth – Holds a doctorate in philosophy from the Université libre de Bruxelles (ULB). He is Associate Professor at the Université Catholique de Lille's ETHICS Laboratory, within the "Ethics, Technologies and Humanities" chair. His research lies between philosophy of technology and critical social theory, and particularly focuses on the normative relationship between machine and human learning in the era of artificial intelligence. With Thomas Berns, he has co-authored *Éthique de la communication et de l'information* (2021). He will soon be publishing *L'intelligence des villes. Critique d'une transparence sans fin* (2023).

Acknowledgments

First of all, we would like to thank the Collège international de philosophie (CIPh; International College of Philosophy) for providing us with the funding for the "Pratiques et usages contemporains des philosophies des techniques" ("Contemporary Practices and Uses of Philosophies of Technology") conference, which took place in Paris in June 2021 and laid the foundations for this book.

This project has received funding from the European Union's Horizon 2020 research and innovation program under Marie Sklodowska-Curie grant agreement no. 945478.

Our thanks also go to the Institute of Philosophy of the Slovak Academy of Sciences, where Élise Lamy-Rested is carrying out her research project co-financed by the European Union.

Lastly, we would like to thank the Laboratoire d'anthropologie politique (UMR 8177 CNRS and EHESS), of which Valérie Charolles is a member, for its support of this long-term project.

Marie Skłodowska-Curie Actions

laboratoire d'anthropologie politique

COLLEGE INTERNATIONAL DE PHILOSOPHIE

Acknowledgments written by Valérie CHAROLLES and Élise LAMY-RESTED.

Introduction

It is typical to assert that our contemporary societies are structured by what we call "digital technologies" as a result of the way they operate through computer codes and the discourses they initiate among the actors that are incorporated into their networks. The growing role played by these new technologies, the profound disruptions they are introducing and our inability to fully understand how they work give rise to increasing concern.

Added to this are the ecological anxieties of our time. The technologies that populate our homes and the technical processes used to extract resources are to be blamed for the ecological emergency we are facing today. Technologies defined in this way only serve a world whose objective is infinite growth at an ever-increasing pace, with total disregard for the Earth's limits, the fragility of living organisms and the poorest countries.

This book takes these two common-sense observations as its starting point, treating them as an undeniable reality. The aim is not only to provide a philosophical understanding of this new situation, but also, and above all, to offer tools for action.

However, what can the philosophies of technologies, with their Greek origins, do in the face of such changes? Are they not too "ancient" to be able to deal with these issues? We have taken a gamble and assumed that they are not, because of the ability of philosophy to reformulate problems, put them into context and reveal what is at stake. This book is the result of a joint

---

Introduction written by Valérie CHAROLLES and Élise LAMY-RESTED.

discussion held in June 2021, under the aegis of the Collège international de philosophie at a conference entitled "Pratiques et usages contemporains des philosophies des techniques" ("Contemporary Practices and Uses of Philosophies of Technology"). The aim was to show that philosophies of technologies are not simply conceptual instruments for making our societies intelligible, but that they are veritable critical weapons with the power to transform the social world.

Why talk about "philosophies of technologies" and why specify "new technologies"? Derived from the Greek word *techné*, meaning "know-how", technologies are originally the arts of making material objects for the craftsman or artist, making speeches for the rhetorician or poet, acting for the comedian or healing for the physician. The Greek *techné* thus refers to the manner of production, in other words, the means the subject uses to achieve an end. Such a definition is based on the idea that only human beings with the capacity for representation are capable of producing "technological" or "artificial" objects, not created by nature. This division between technology and nature structured a large part of Western philosophy right up to Descartes, who understood technologies as instruments that enabled human beings to design and build machines to "render themselves as masters and possessors of nature", as he wrote in his *Discourse on Method*. Our entry into the modern era thus coincides with the technological project to dominate nature through the art of making mechanically operated tools and the power to control its forces.

This is undoubtedly the definition we refer to today when we distinguish technologies from new technologies, which appear more fluid, more abstract and more incomprehensible because their mechanisms remain hidden to those who are not computer experts. Therefore, not only are new technologies perceived as immaterial, they are also shrouded in a kind of mystery, the secret of which is known only to a select few. For all these reasons, they no longer have anything to do with the "good old" materiality of the locomotive or even the pendulum.

However, it is precisely because we want to underline the materiality of new technologies that we have included them in the term "technologies". A computer does not work without electricity; it is made up of materials that we take from resources in the Earth, and its software is the result of the labor of engineers. The Internet is not simply a network of dematerialized information connecting all parts of the world at the speed of light. In order to function, it consumes electricity and batteries are made from rare and expensive metals. As for the data collected, most of which has been naively entrusted to the networks, it is kept in huge physical storage spaces that are expensive to maintain. In other words, the cloud is an illusion; the "Internet cloud" could, ironically, instead be the result of the CO_2 emissions generated by its operation.

Regardless of what is assumed, all so-called immaterial inventions, from Pascal's calculator to supposed "artificial intelligence", remain technologies invented by an irreducibly mortal organism. The abstractive capacities of computing machines entirely depend on the production of a destructible object. Even though technologies are diverse and their foundations can be considered in equally diverse ways, they nonetheless involve objects made up of materials that will be extracted from planet Earth until they are completely exhausted. Having reached this stage, perhaps we will seek out these materials in space, a hypothesis that is attracting the attention of Google (Alphabet), Apple, Facebook (Meta), Amazon and Microsoft (GAFAM). At a metaphysical level, we cannot ignore this reality, nor can we assert that the supposedly infinite power of technologies would be an ontological fact that cannot be disputed.

From this perspective, new technologies are just one form of technology, the latter being understood, through its Greek meaning, as ways of producing tools that enable living organisms to adapt to or modify their environment. This book thus adopts a broad definition of technologies as tools invented and developed by human societies to interact in a singular way with their environment. This environment is therefore also technological, and we can even hypothesize that the practices of technologies are not reserved for humans in the living world. There is no intention for this book to fall into a "technophobia", promoting an unrealistic return to a nature that is, by definition, non-technological (Lamy-Rested 2022)[1].

Our book is rooted in a shared concern: how can the philosophies of technologies make contemporary times intelligible, and give us new means for action? To answer this question, which involves the future of our

1 Lamy-Rested, É. (2022). The technical object at its limits: Derrida reader of Husserl. *Philosophy and Technology* [Online]. Available at: https://link.springer.com/article/10.1007/s13347-022-00513-7.

societies, we have chosen to interweave the voices of philosophers from different backgrounds. Based on the idea that the philosophies of *techné*, which were invented in ancient Greece, have more to say than ever about the Internet, "artificial intelligence" or the problems generated by the over-intensive use of technologies, we have chosen not to ignore any philosophical tradition, and instead bring them into the conversation.

This polyphonic work is an unprecedented attempt to disconnect technologies from the logics of power that endanger life. Its 13 contributors address themes that are particularly critical today:

– *Continuities and disruptions in the practices of philosophies of technologies*: What philosophical legacies can we draw on? Is the reflection on technologies that has been going on since Antiquity still relevant today in a context where the technical fact has become so massive?

– *The epistemological issues of contemporary technology*: What scientific models should be put forward? In particular, how can we analyze the strength and fragility of tools forged using technologies? How can we deal with the misunderstandings and misconceptions surrounding the use of code-based tools?

– *The subject in the era of digital metamorphosis*: In what ways do technologies currently affect the constitution of the individual, creating new capacities and alienations? Can thought be reduced to computer calculating power? If not, what is the link between the machine and the subject?

– *Testing politics through technologies*: What is the role of politics in the face of new surveillance tools and the limitations on democratic capacities that originate from a poor understanding of technologies? How can we unleash the collective capacity to act that technologies give us and the power we have over them? What is the significance of democratic decision-making when it comes to technologies?

Throughout this book, the aim is to take a serious look at a movement taking place in the contemporary world. The level of development of our technologies, and their dissemination across the whole range of our activities, marks a turning point. In less than a century, technologies have acquired unprecedented power in all fields, from the economy and knowledge production to decision-making and interpersonal relations. The result is a world with a new form: the "infinite universe" of classical science, to which humans and their free will were opposed, has given way to a "reflexive system". Like social networks and financial markets, this new paradigm operates through echoing and mirroring mechanisms and bears the mark of interdependence between humans, technologies and ecosystems. It signals the end of naturality, in the sense that there is no longer any place on Earth, including the seabed, that has not been transformed by the products of our technologies (Charolles 2013)[2]. This mutation is not only challenging our forms of life – intimate, economic, social and political – but also our frames of thought. This book draws the consequences.

In order to make our way through this kaleidoscopic system, and to distinguish what is a crisis and what is not, we need suitable conceptual frameworks. Faced with technologies that are increasingly difficult to understand in their modes of operation, and that evolve very rapidly and intertwine with one another, our capacities for understanding and orientation are indeed being tested. They are tested first and foremost by the language of computer code, a language that has become inescapable, but in which very few people are literate, rendering all others in a position of illiteracy. This is the breeding ground for many misunderstandings, and we risk deluding ourselves on themes such as "artificial intelligence", the possibility of objectivity or a sense of time that is inextricably linked to progress. Through its diverse themes, this book aims to help everyone understand the technical facts at work today.

It also aims to provide guidelines on the conceptual frameworks that can enable us to grasp reality, where the partition between that which is natural and that which is artificial no longer makes sense, if it ever did. In such a context, it is no longer possible to rely on the conceptual framework of the Enlightenment based on the opposition between man and nature or between morality and calculation. We need different categories to orientate ourselves effectively in the contemporary world. Indeed, our technical capabilities are reshaping the way we inhabit space (which is no longer merely a physical distance, but also a capacity for connection) and time (the regularity of movement in Newtonian mechanics is giving way to far more random and extreme modes of operation). This new paradigm brings to light new forms of relationships that are far more refined than the determinism of an outdated vision of physical science, which does not allow for individual or collective decisions. This book provides a set of keys in this regard.

2 Charolles, V. (2013). *Philosophie de l'écran*. Fayard, Paris.

By reminding us that technologies have always been part of human life and that, however sophisticated, they are the result of human labor, philosophy ultimately allows us to mourn the loss of technical determinism and realize that there is no need for our tools to be deployed in one direction and not another. This is a common point throughout the whole book: assuming a shared contingency in the deployment of technologies creates spaces in which politics, as the voice of citizens, can intervene with legitimacy and success. As the chapters unfold, we see the extent to which technologies are embedded in social and political aspects. They are conceived with a view to certain social forms and are not immune to power relations, which are particularly powerful in the economic sphere today. In a democracy that respects the expressed will of enlightened citizens, the question of how technical deployments are conceived deserves to be taken seriously, as do the responses of the population. There is nothing inevitable about technologies' reliance on power at an "ever increasing" pace that is unsustainable for living organisms and ecosystems. In fact, it goes against the grain of what the ancient nations intended for *techné* and could rightly give way to other prospects that are more respectful of people's diversity and free will.

This book speculates that philosophies are capable of responding to technical changes that can also be seen as opportunities to shape our future differently. Philosophies' diverse practices and uses converge to demonstrate the power that technologies offer us, as well as the power we have over them.

Faced with the sprawling, metamorphic deployment of digital technologies and the resulting loss of points of reference, we need a philosophical moment.

Part 1

Continuities and Disruptions in the Practices of Philosophies of Technologies

Introduction to Part 1

After ancient history, when the term *techné* had a place in thought, the latter made a comeback with Marx, who was undoubtedly the first philosopher to question the metaphysical foundations of technical production, not just to reveal them, but to denounce their possible disruptive drifts. It was not until the 1930s that philosophy – or psychoanalysis – began to sense technology's role in the possible destruction of the world. Husserl's *Krisis*, Bergson's *The Two Sources of Morality and Religion* and Freud's *Civilization and Its Discontents* are highly representative in this respect. Written at a time when the First World War, which was also the first technological war, was still fresh in people's minds and when the Nazis' accession to power raised the possibility of another massively destructive war, these three books examine the meaning and role of technology, which is identified through mechanical processes that serve the desire – the drive, as Freud would say – for omnipotence.

Today, we face an unprecedented issue. The multi-dimensional ecological crisis we are experiencing is forcing philosophy to revisit the metaphysical foundations of technical production in an unprecedented way. It is now only based on the limited finite Earth, whose resources are progressively being depleted, endangering all living things, that the philosophies of technology, from their Greek origins onwards, must be reread, questioned and practiced. As Marx already pointed out in more specifically economic terms, we can no longer read philosophy as a mere instrument for understanding and contemplating the world, but must practice it in such a way that it also allows us to grasp it. The first three chapters of this book – Pierre Caye's "The

---

Introduction written by Élise LAMY-RESTED.

Question of Technology and Ecological Constraints", Xavier Guchet's "From Power to Care: For an Object-Oriented Philosophy of Technology" and Élise Lamy-Rested's "Thinking in the Anthropocene Era with Henri Bergson" – show how philosophies of technology, regardless of their era, can become critical weapons of our modern day, helping us to completely change our relationship with technology.

The first chapter, by Pierre Caye, takes a new look at the metaphysical foundation of technology "in the face of ecological constraint", which is first and foremost a thought about limits, and proposes an ecological metaphysics. While philosophy may appear to link technology to power and set aside the question of limits, Pierre Caye reverses this vision, starting again with Proclus' philosophy. Together with Heidegger, the author rereads Proclus' commentary on Plato's *Timaeus*, and sees it as an invitation to think of technology in terms of the limitation of space and time, rather than in terms of the *hubris* of omnipotence.

By re-reading the various interpretations of the Prometheus myth, Xavier Guchet echoes this reflection, then shows how *techné*, in its Greek origins, is not connected to power and *hubris* but to care. He encourages us to reorient it in a pragmatic and practical way. Far from the expertise that produces fixed objects that have a predetermined place and role in the world (e.g. the shoemaker's sole function is to make shoes), technology actually invents malleable objects. Technology is part of a set of references that embrace the change and movement characteristic of living beings, from which technical objects are not, ultimately, so distinct. The purpose of these objects is not to increase our power, but above all to help those that are alive to be better off.

Lastly, Élise Lamy-Rested follows this ecological, pragmatic and materialist approach based on Bergson's philosophy of technology, which does not distinguish technology from living or organic beings. As a result of them being products of life and not just of intelligence, technology is ontologically inscribed within the limits of the living. It contributes to life's protection and flourishing, which has little to do with power. Through an explicit reading of *L'évolution créatrice*, Élise Lamy-Rested argues that Bergson's philosophy of technology can help us extricate ourselves, at least intellectually, psychologically and politically, from the Anthropocene era, which is traditionally defined as the new geological period resulting from human action on the Earth, conceived as a system.

1

The Question of Technology and Ecological Constraints

Today, we can speak of the Western technology crisis, just as Husserl spoke of the "European sciences crisis" almost a century ago, with the same paradoxical spirit. At the end of his life in the mid-1930s, when Husserl was writing the texts that would make up his final work, the famous posthumous collection on *The Crisis of the European Sciences and Transcendental Phenomenology* (Husserl 1954), the sciences had never experienced such a renewal; not only the physical sciences, which were shaken up by the concurrent double revolution of general relativity and quantum mechanics, but also the human sciences, which were then being devoted to all areas of human life. It seems no less paradoxical to speak of a technology crisis today. Technology has never been so developed or so inventive. The digital revolution and new information and communication techniques have profoundly revolutionized our behavior, our modes of production and even our modes of government; nanotechnologies and biotechnologies, in turn, seem to be taking over, promising equally spectacular revolutions. However, therein lies the paradox: over the past 30 years, we have been going through a phase of profound technological changes, yet we have not pushed past the boundaries of shortages, lack and scarcity[1]. Famines are far from being eradicated; resources are running out, appearing to be less and less renewable; and the struggle for the appropriation of goods is proving increasingly fierce both within nations and between them: oil wars, water

Chapter written by Pierre CAYE.

1 Per capita food consumption peaked in 1984 and has been falling ever since.

wars, wheat wars and the list of these kinds of conflicts continues to grow[2]. The ecological situation is becoming more and more obvious, and some people are calling for growth, or even degrowth, to compensate for this state of affairs. We are increasingly witnessing a sort of decorrelation between technological invention, growth, abundance and human satisfaction to the point where a growing number of people are calling for a revolution, not so much in technology (for these revolutions are already happening), but in modes of production, under the admittedly vague and ambiguous term "sustainable development".

1.1. What is the appropriate metaphysics for ecology?

When discussing the epistemological situation of the sciences between the two world wars, Husserl said very little about their links with technologies, even though this soon became a central issue for philosophers (Heidegger) and politicians alike after the war (the founding of the Centre National de la Recherche Scientifique in France). In a way, the increasingly close intertwining of sciences and technologies is a response to Husserl's concern, reintegrating science into the field of "life", even if *life* here is thought of in terms of use and power, rather than reason and humanity's fulfillment in transcendental subjectivity. That being said, Husserl would have had to explain how Western reason and its philosophical destiny, which he so often evokes in his reflections, remain totally foreign to the desire for power expressed in the technological becoming of metaphysics.

Just as the excess of reason in scientific positivism seemed, in Husserl's eyes, to render the sciences of his time unsuitable for human progress and rationality, the overwhelming power of scientific-technological devices seems, in turn, to exceed human life and transform it into a death drive.

Regardless of the changes in epistemological questioning brought about by the transition from the scientific positivism of the 19th and early 20th centuries to the modern day's scientific-technological devices, the fact remains that reference to Husserl's testament is still relevant. Indeed, it

2 The slowdown in overall productivity is occurring at the same time that technical innovations are multiplying and R&D expenditure, as a percentage of GDP, is rising. This is the productivity paradox, a paradox summed up by Nobel Prize-winning economist Robert Solow: "You can see the computer age everywhere but in the productivity statistics" (*New York Times Book Review*, July 1987).

seems that the aporias of the contemporary scientific-technological devices generated by the future of Western rationality can only be overcome by a renewed effort that stems from reason and not by its denunciation. The birth and revival of dialectical reasoning will determine what is at stake in the modern era, the future capacity of which we are trying to fathom in the face of everything that is likely to call it into question.

However, we must also be aware of the difficulty of exclusively relying on Husserl when approaching the question of dialectical reasoning in the face of contemporary technologies, precisely because we come up against a question that was beginning to emerge in Husserl's time, but which had not yet been integrated into philosophical discourse, namely the ecological constraint that doubly challenges the Husserlian gesture:

– It questions the Husserlian definition regarding the origin of science as the infinite potential of thought, as the infinitization of humanity's intellectual potentialities, which is not enough to overcome the aporias of the contemporary productive system, but instead accentuates them, thus testifying to the contemporary theoretical and practical crisis of infinitism.

– This leads to a second challenge, that of the nature of scientific-technological rationality, which must be overcome by reason itself: it is not only positivist rationality, instrumental reason and engineering reason that ecology challenges, but, more radically, metaphysical reason, and particularly idealist metaphysics, the very dialectic of Being and reason, in other words, the metaphysical foundations of science. This means that the dialectic of technological reason that we are trying to think through will appear first and foremost as a critique of dialectical reasoning, as established by the systematics of German idealism.

We call it an ecological "constraint" because it is not a question of theorizing about ecology, but simply of noting what the ecological question implies for thought. The ecological observation is simple and obvious. Ecology notes the finite nature of the Earth, the limits of its depths, the depletion of its resources and the extent of their impossible renewal. The value of this approach is that it points out a double paradox that severely impedes our relationship with reality:

– The infinity paradox: the more possibilities our technological inventions allow us to open up in our actions, the more we discover their limits, in other words, the more we experience the finiteness of the world, its exhaustibility and non-regenerative nature, which in turn marks the limits of our condition. Older generations lamented their powerlessness in the face of matter's resistance and prominence; today, we suffer from the powerlessness to avoid matter's depletion: but, in both cases, the substance is disappearing and is beyond our grasp.

– The power paradox: the more power – particularly technological power – we wield, the more we feel powerless in the face of what we are doing, that is to say, the more we generate adverse effects that we can neither treat nor foresee, according to the very principles of complexity theory.

However, we cannot ignore idealism's objection to the empirical findings of ecology. For idealism, matter is finite, but the mind is not. What matter cannot do, the mind can. In other words, human invention and ingenuity are always capable of restoring the Earth's abundance: the philosophical roots of what is now known as neo-productivism. In the eyes of the consistent idealist, ecology thus simply appears as the doxa of a productive and technological civilization on its last legs – ours – with diminishing returns, where the slightest point of growth is paid for by ever more numerous and costly negative externalities (pollution, the increasing scarcity of raw materials and hence their increasing cost, reinforcing the alienation of humans and their exploitation by humans), which can be seen as a regime awaiting a new production system that is capable of providing the material and technological conditions for a new society of abundance.

Today, this technological messianism no longer works because there has been a hiatus between the eschatological promise of the idealist dialectic and the logic of its operation from the outset. Before highlighting this hiatus, it is important to emphasize the importance of the question of the Hegelian dialectic and its multiple avatars, in the sense that this dialectic is first and foremost the thought of work, or more precisely, thought in work, insofar as *work* here is the source of and condition for the transformation of the world, which points to a conception of metaphysics as the radical thought of the productivity of Being. Marx writes the following of Hegel:

> The importance of Hegel's *Phenomenology* and its final result – the dialectic of negativity as the moving and producing principle – lies in the fact that Hegel conceives the self-creation of man as a process, objectification as loss of object [*Entgegenstandlichung*], as alienation and as supersession of this alienation; that he therefore grasps the nature of labor and

> conceives objective man – true, because real man – as the result of his own labor (Marx 1932, p. 156).

The promise of the idealist dialectic is the total assumption of substance by the mind. The mind reinvigorates the substance on the verge of exhaustion, while the renewed matter in turn provides the mind with richer and more powerful resources, enabling it to carry out ever more audacious projects, in a kind of ladderlike manner between mind and substance, until matter is nothing but infinite energy or light, and as such totally identifiable with the mind, according to the principle of relief and assumption.

However, in terms of the effectiveness of its functioning – and the effectiveness of its functioning must be understood not only as its functioning in reality, but in how this functioning produces reality – the idealist dialectic does not go that far: it remains asymptotic.

In dialectical terms, the asymptote can be interpreted in two ways: either it marks the end of history, the increasingly horizontal evolution of society, or the flat encephalogram, according to the traditional thesis defended by Kojève and taken up by Francis Fukuyama; but in reality, by virtue of the impossibility of which it is the result and the ever-uncompleted action of attraction that moves it, the asymptote enters into random phases of discontinuity, which generate chaos-like vibrations that are typical of the limit states that bring about phase transitions, with these vibrations being all the stronger the longer the phase transition is delayed. The dialectic pushes mutual intensification to the limit without ever identifying and converting the substance and mind to the absolute presence of Being, thus imposing a situation of unresolved tension on the motion of Being. It is precisely this asymptotic situation, with no possibility of assumption, of perfect convertibility between mind and substance, between will and Being, this perpetual source without sabbath that is always at work, that exhausts the substance and necessarily leads to its decay. This decay must be met with something other than a simple promise, even if it is the ideal promise. To put it another way, although it is true that dialectics serves to explain the articulation of Being, becoming and power, as well as to explain the processes of convertibility, the dialectical situation, as we have just described it, leaves Being, becoming and power in a state of inarticulation or subject to the impossibility of articulation.

In other words, the European science and technology crisis is essentially a crisis of ontology, the process of Being, of Being as a process, and thus of dialectics in the metaphysical sense. We are therefore a long way from the Husserlian diagnosis. However, the point here is not so much to save dialectics and put it back on its feet, but to assume this inarticulation, to think of becoming without immediately identifying it with the omnipotence of Being, to think of a force that is not necessarily the formation of Being and the dynamic of becoming or in the process of becoming, to think of Being as something other than the formation and dispensation of power.

This is why we will be taking a completely different path from the Husserlian diagnosis in order to think of technical rationality in an equal sense to the fundamental aporias we have just described. To this end, we propose to meditate and comment on Proclus' commentary on Plato's *Timaeus*, and in particular on one sentence, a thought that is particularly significant for the modern day's question regarding technology: "The mixture [...] of technology (*to technicon*) and self-development [of nature] (*to autophues*) is like the mixture that takes place in divine creation (*hê theia poiêsis*), between the limit (*ho horos*) that comes from itself and the procession (*to proïon*) that comes from Being (*to einai*) and substance (*hê ousia*)" (Proclus 1965, p. 60).

1.2. Technology and limits

Neo-Platonism, of which Proclus is one of the most emblematic representations, does more than simply revisit and comment on Plato's work. It is, in fact, the most accomplished summary of over eight centuries of ancient philosophy between Plato, Aristotle and the Stoics. This summary, which deserves to be better known and studied, forms the matrix of Western philosophy right up to the present day. Neo-Platonism, which is strongly imbued with lessons from Plato's *Parmenides*, which make the latter appear as a systematization, radically distinguishes Being and the One, as *Parmenides'* first hypothesis teaches, to the point of suspending their convertibility, which is nevertheless the most constant feature of most of the ancient philosophical schools that preceded it. This unprecedented and singular break in the history of philosophy is itself a source of ambiguity that is precisely what makes philosophy, which at first glance appears to be pure dogmatics, so interesting and topical. This prevalence of the One seems to justify the ontotheological nature of metaphysics (which conceives the totality of reality as a derivation of the principle), if not its ontotheological-political nature (which identifies the principle with power and its dispensation to the world). As such, neo-Platonism is undoubtedly at the origin of the assimilation of metaphysics to power and reign, its technological becoming and identification with the productive destiny of the West, according to Heidegger's genealogy of metaphysics.

However, the radical difference between the One and Being does not necessarily lead to the oblivion of Being. We must not downplay the radically critical dimension for metaphysics at stake in the non-convertibility of the One and Being, which also leads us to meditate on the principle's non-application, on its impossibility of proceeding, and consequently, if not on its powerlessness, at least on its unproductivity. This ambiguity is also present in Proclus, who clearly separates power from principle, and in Damascius, who distinguishes two moments in principality: the "One of the All before the All" endowed with all the attributes of omnipotence, opposed by the supreme, ineffable One, which radically exceeds not only the whole, but its very principle, and which consequently takes leave of all totalization and derivation.

Proclus' sentence is part of the counter-movement that neo-Platonism is behind. It is only one stage in the process, given that the *Timaeus*, which until the 12th century served as a fundamental reference for the philosophy of nature before being supplanted by Aristotelian physics, is, in the eyes of neo-Platonism itself and in the best conditions of understanding, the ideal propaedeutic for approaching the difficulties of the relationship between Being and the One, as formulated in the hypotheses of *Parmenides*.

Proclus thus notes that the demiurge's creation is a mixture of the artificial and the natural, or, more precisely, technology (*to technikon*) and the self-development of *phusis* (*to autophues*), in the image of divine creation (*hê poiêsis theia*), which is a mixture between the limit that comes from itself (*ho éph'eautês horos*) and the procession that comes from Being and substance (*to apo tou einai kai tês ousias proïon*).

It is clear that this line of thought, which radically and fundamentally questions the essence of technology, remains outside the question of technologies as formulated in the 19th and 20th centuries, even though it is clear that something important that concerns us is being said, and which, in the context of the contemporary productive system, has never been said before.

What is *autophusis*? It is *phusis* insofar as it proceeds from itself and unfolds by itself. This simple term says something about neo-Platonism that is not usually heard, namely that this metaphysics, which appears as the pinnacle of transcendental metaphysics, is initially a thought of *phusis*, and even more so of the self-procession of *phusis*, that is to say, of its autonomy. While there is certainly a first philosophy in neo-Platonism, that of the first principle, the latter is inevitably led to respect the autonomy of *phusis*, and consequently cannot play a foundational or determining role.

However, the question of technology, as posed in the 20th century by both the Heideggerian and Bergsonian traditions – the two dominant interpretive traditions – only fits and makes sense within the framework of this self-procession of nature, postulating the absorption of *technikon* into *autophues*. Technology contributes to the procession of *phusis* and to its unveiling when it is assimilated and naturalized to the point where the boundary between art and nature, the given and the constructed, blurs and merges in the self-procession of Being.

This absorption of technology into the self-procession of Being can be interpreted in two different ways.

Either we consider that technology contributes to the intensification of self-procession, according to the Bergsonian interpretation taken up by Leroi-Gourhan and Simondon, which pictures an increasingly systemic and concrete development of the technological object, to the point of its naturalization according to Simondon's thesis.

Or, on the contrary, we consider that technology, in its rational techno-scientific dimension, leads to the blocking of procession, its alienation and its exhaustion. This is what Heidegger calls the *Gestell*, the "device", a term that signifies the systemic and processual nature of the world of technology, but also the petrification and stagnation of what "is questioned". That said, Heidegger's task is not to dissociate technology and Being, but to think, by returning to Aristotelian physics in his commentary on the first chapter of Book II of Aristotle's *Physics*: "Vom Wesen und Begriff der Φὐσις: Aristoteles, Physik B, 1" (Heidegger 1976). This could just as well be entitled "Vom Wesen und Begriff der Τἑχνη", or better yet, "What does Aristotle mean by production (*Herstellen*)?", which could be a *tékhnê* that is in tune with nature, and as such contributes to the freedom of the system, in other words, to the unblocking of the articulation of humanity, world and principle, or to the intensification of Being, becoming, power and their convertibility[3].

What neo-Platonism proposes in Proclus' phrase is something quite different: a clear distinction between *phusis* and *technikon*. One might object: is this not returning to the old state of the issue, to viewing technology as a mere means to serve humans, in an external position to nature, which they seek to master and appropriate? We would thus be returning to an exterior technology, to the Cartesian interpretation of science and its methodical project, which is supposed to make humanity the master and possessor of nature. Yet is it not this very exteriority that explains why, in Heidegger's view, technology blocks the procession rather than promotes it? In other words, is Proclus' phrase not just the metaphysical starting point that necessarily ends in the *Gestell* of the contemporary productive system? Under these conditions, would it not be a dangerous misunderstanding to use this phrase as the basis for a fundamental meditation on sustainable development? We are well aware of these important objections, which lead to us having a precarious explanation of Proclus' phrase. In this respect, two preliminary remarks are called for.

First of all, it is important to highlight the strange nature of Proclus' definition in relation to our usual criteria: here, technology is not at the service of production, its growth and intensification, but aims instead to limit its expansion, as if intensification were not the sole purpose of technology, as if technology and intensification did not belong together, or as if the question of intensification is downstream from that of technology and says nothing in itself about the origin and essence of the latter. Through its limiting and delimiting function, technology has a purpose other than production; in Proclusian vocabulary, it is a matter of *phroura*, in other words, the guarding, maintenance and conservation of nature in its process[4].

---

3 "The τέχνη can only go as far as the φύσις [...] [Die τέχνη kann der phusis nur entgegenkommen [...]" (Heidegger 1976, p. 257). On the question of production and its relation to *phusis*: *ibid*., p. 281; French translation, p. 241.

4 On the *phroura*, see Proclus (1968), p. 123; see also Proclus (1987), p. 32 (*hê phroura atreptos*: "the immovable guard"); on the expression *hé aklinês phulakê*, "the inflexible guard", see Proclus (1987), p. 125. On the importance of the notion of *sôtéria*, both salvation and preservation, in Plato, see Sandvoss (1971). The term *sôzein* also recurs repeatedly in Proclus' Platonic theology. See in particular Proclus (1968), p. 102, where the term *sôzein* is related to the term *phulattein*, "to stand guard".

Although technology here shows its strength (and indeed, it is not a question of succumbing to the charms of idylls and "sheepfolds": there can be no technology without force, without raising the question of its efficiency), this force is not measured in energy or power, in *energeia* or *dunamis*, but what Proclus calls *sunektikon*[5], in other words, a force of gathering, of coherence of Being that protects nature's process, its self-deployment, from discord, chaos and disintegration.

In the genealogy of technology, there is therefore a moment of bifurcation, with one branch going in the direction of intensification, and the other toward guarding, so that any thought questioning the essence of technology is driven upstream of this bifurcation.

This first remark in no way resolves the major objection we formulated earlier, namely that, as a limit, technology stifles the deployment of *phusis*, an objection that renders the very purpose of guarding meaningless, which obviously only makes sense insofar as it serves life, and not its exhaustion.

Resolving this objection requires a better understanding of the limit and its significance. As the metaphysics of the non-convertibility between the One and the Being allows us to do, we need to step outside the framework of all ontological poetics, whether it be hylemorphic or morphogenetic: here, the limit is not the form that configures matter, like a seal on wax (hylemorphic model), nor the final stasis that determines the end of a process, balancing its forces and solidifying its result (morphogenesis). If we stick to ontological definitions of the limit, to the function of the limit, in a purely ontological framework, characterized by the interaction between form and matter or by that of act and power, the *technikon* would effectively appear as a corset leading to the *Gestell*. As the *technikon* leaves the ontological field, it not only dissociates itself from any intensification function, but even more radically from any production function.

5 Regarding "that-which-holds-together" (*to sunektikon*), see Proclus (1974), p. 18. Proclus speaks in III, 8 (Proclus 1978, p. 32) of the divinity that holds together (*hê théotês sunektiké*) or of the goodness that holds together (*hê agathotês sunektikê*) (Proclus 1968, p. 91), which Proclus distinguishes from the goodness that accomplishes and perfects (*hé agathotês télésiourgos*) and the goodness that gathers (*hê agathotês sunagôgos*). On the close relationship between guard (*hê phroura*) and synectics, see the terms *phrourétika kai sunektika* or *to sunechein kai to phrourein* (Proclus 1968, p. 123).

Production, *producere*, means "to drive forward", to promote the progress of Being, its coming into presence (*Anwesung*), its blossoming (*Unverborgenheit des Seienden*), as Heidegger would say. In a way, Heidegger's dispute over the essence of technology does not go beyond the realm of production. He focuses on the authenticity of the latter, its autonomy (*to autophues*), the nature of its deployment, thus distinguishing between production identified with provocation, with the subservient mobilization of nature by humans to which modern technology leads, and the free favor of Being that flourishes under the influence of Heidegger's *téchnê*[6].

In turn, the Proclusian *technikon* withdraws from the field of production; it dissociates *poiêsis* from production and *theia poiêsis* from creation. It is about withdrawal and restraint, not progress. *Ho eph'eautês horos* is first and foremost the limit that *poiêsis* imposes on itself, before being the limit it imposes on the procession of Being.

Withdrawal can have three meanings:

– a mystical meaning that implies withdrawing from the whole of reality by way of excess, and we do not mean excess from a moral point of view, but purely metaphysical, in other words, that which exceeds all reality. As excess, this withdrawal is, as we understand it, foreign to the *technikon*, both because excess as such is limitless, but also because it has no effect on reality;

– an ascetic meaning that corresponds to a certain topicality of ecology. We understand asceticism here as the desire to reduce the impact of human activities on the Earth at the cost of degrowth. Here, we find a certain Stoic spirit, according to which we must conform our desires to the state of the world, given that we cannot subject the world to our desires. Morality then takes precedence over the question of technology, so that the ascetic approach does not allow us to think about the *technikon*'s own operability in its guard function; in other words, it does not require us to rethink the technological paradigm, but is content with implementing the political and

6 "We still by no means think decisively enough about the essence of action. One knows action only as the bringing about of an effect, the effectiveness of which is assessed according to its usefulness. However, the essence of action is perfecting. Perfecting means to unfold something in the fullness of its essence, to bring it forth; *producere*" ("Wir bedenken das Wesen des Handelns noch lange nicht enschieden genug. Man kennt das Handeln nur als das Bewirken einer Wirkung. Deren Wirklichkeit wird geschätzt nach ihrem Nutzen. Aber das Wesen das Handelns ist das Vollbringen. Vollbringen heißt : etwas in die Fülle seines Wesens entfalten, in diese hervorgeleiten, producer") (Heidegger 1947, p. 5).

moral conditions for the decommissioning and exhaustion of the current paradigm; the decreasing approach does not imply a transformation of the productive system, but simply the reduction of its impact. From an economic point of view, this belief that degrowth will lead to a change in production methods is akin to an ascetic approach to ecology;

– we understand withdrawal in a third way, a withdrawal that, while extricating us from the ontological domain of reality and its production, nonetheless keeps us within the field of reality's constitution. We withdraw from Being, so that we can access the conditions of its presentification, in other words, space and time. The delimiting function of technology is not to do with matter and Being, but with space and time: this is the fundamental thesis we are defending, which is, as we understand it, different from the strictly neo-Platonic doctrine[7], but which is likely to give it a new functioning for the time being. If the question of the original rationality of technology has any relevance, it is at this level that it must be presented. Better still, if the origin of technology is defined in terms of rationality rather than the desire for power, it is precisely because it is not at the level of the emergence of Being, but instead at the level of the delimitation of space and time.

We can only modify the technological paradigm by thinking of its effectiveness at the level of space and time, while making it clear that at this level, the meaning of the effectiveness of technology is radically different from the kind of ontological effectiveness demanded by production, be it "provocation" or "blooming", total mobilization or "letting it be" (*Gelassenheit*). Indeed, it is precisely because technology operates on space and time that its force is *sunektikon*, not *energeia* or *dunamis*, and its efficiency contributes to the coherence of Being, not its intensification.

1.3. For transcendental poetics: technology at the service of our relationship with space and time

This definition of technology as withdrawal rather than production changes the very meaning of the limit. The limit does not inform objects nor configure a world; it is instead about differentiation, mediation, separation,

7 Although it would be possible to show the extent to which the categories of time and space, particularly in Damascius, play an important role in the operation of the *sunektikon*, see Joseph Combès' remarks on Damascius' lost treatise, *Du nombre, du lieu et du temps* (see Galperine 1980, pp. 325–341; Damascius 2002, pp. xli–xlvi).

distancing, spacing and expansion, without which no "human" world could be established, the world being thought of here as an ark and a protection, not as a machine or a *fabrica*. This expansion involves both space and time.

The expansion of space is all the more necessary as the combination of the demographic explosion and the speed of communications tends, on the contrary, to contract it. There are two ways of thinking about this expansion: in a purely quantitative, extensive way, but also in a productivist way, which is reflected in the desire to conquer space and in dreams of terraforming other planets. However, for a long time to come, Earth will most likely remain our only vital space, our soil (*Boden*) and our ark (*Arche*, *Erde*), in the words of Husserl (Husserl 1940, pp. 317 and 321). Under these conditions, technology's task is to increase livable indoor space in a qualitative and discreet manner. It is therefore hardly surprising that questions about spatial planning, urbanism and architecture account for more than half of the literature on sustainable development.

Spacing and expansion also involve time. Technology is originally linked to the construction of time, contributing to humanity's foresight and anticipation of the future. It goes without saying that Prometheus, the mythological father of human technology, represents foresight, knowledge that projects itself (*Pro-mathésis*) and, by projecting itself, builds time and masters the future. More historically, and less mythologically, we know to what extent the constitution of modern technology from the *Quattrocento* onwards is linked to the birth of the art project and the development of its design logics. That said, complexity theories have taught us that time cannot be controlled, that our plans are constantly thwarted, and that our desire to control the future generates adverse effects that turn against us.

Under these conditions, we must ask ourselves whether there is a technology capable of expanding time and building the future without Prometheism, in other words, without the hope of projecting time in order to better control it, or whether we must resort to a non-technological moral and political temporalization, as Plato himself recommends when he relates the myth of Prometheus in the *Protagoras* (322b), where, at the end of Platonic reflection, politics comes to make up for the shortcomings and flaws of technology.

In his short essay *De brevitate vitae*, the first successful attempt at temporalization in the history of philosophy, in other words, constructing time for humanity, Seneca distinguishes between the *dilatio* of time, which slips through our fingers like water from a stream, running ceaselessly, flowing and rushing (*in cursu enim semper est, fluit et praecipitur*), a time at the limit of existence, ceasing before it even comes to pass (*ante desinit esse quam venit*), a time of flight, a flight of time (*fuga temporis*) in its tireless mobility (*irrequieta agitatio*), and ultimately our escape from ourselves in time (X, 6); and *dilatatio* (VI, 4), in other words, its prolongation, the constitution of the presence of time, of time as a presence of oneself to oneself and, as such, the constitution of the present, a densified and prolonged temporal stasis, repelling the illusions of the past and the future produced by memory or hope, through which duration is elaborated, brick by brick, stasis by stasis, the very thing to which all that claims to be sustainable, especially human development, must refer.

Today, technology faces a temporal challenge: how can we manage the expansion of time, give time to time, while meeting our responsibility toward future generations, without still being able to claim to master time or even project ourselves into it? Also, how can we gain or defer time in the continuous acceleration of reality, without falling into the tardiness and regression that the new ideologies of degrowth condemn us to? With these questions alone, we have barely scratched the surface of this thought challenge that Proclus presented over 15 centuries ago under the term *technikon*.

We can only observe the madness affecting our current relationship with technology: technology is an absolute that is infinitely self-produced within the framework of the *autophusis* that it helps to intensify, an absolute that humans only externally limit in the form of some professional and social ethical code, whereas, on the contrary, we need to think, as Proclus invites us to do, of the essential co-ownership of technology and limits, and better still, of limits as the very essence of technology.

1.4. References

Damascius (2002). *Traité des premiers principes, I.* Les Belles Lettres, Paris.

Galperine, M.-P. (1980). Le temps intégral selon Damascius. *Les études philosophiques*, 3, 325–341.

Heidegger, M. (1947). *Letter on Humanism.* Groth, M. (trans.) [Online]. Available at: https://wagner.edu/psychology/files/2013/01/Heidegger-Letter-On-Humanism-Translation-GROTH.pdf.

Heidegger, M. (1968). *Vom Wesen und Begriff der φύσις: Aristoteles, Physik B 1. Wegmarken, Gesamtausgabe, 1. Abteilung: Veröffentlichte schriften 1914-1970.* Vittorio Klostermann, Frankfurt.

Heidegger, M. (1976). Ce qu'est et comment se détermine la φύσις. In *Questions II*, Fédier, F. (trans.). Gallimard, Paris.

Husserl, E. (1940). Grundlegende untersuchungen zum Phänomelogische Ursprung der Räumlichtkeit der Natur. In *Philosophical Essays in Memory of Edmund Husserl*, Farber, M. (ed.). Harvard University Press, Cambridge.

Husserl, E. (1954). *Die Krisis der Europaischen Wissenschaften und die Transzendentale Phaenomologie.* V.Z.W. Husserl-Archief te Leuven and Martinus Nijhoff, La Haye.

Marx, K. (1932). Ébauche d'une critique de l'économie politique. In *Œuvres, Économie II*, Rubel, M. (ed.), Malaquais, J., Orsoni, C. (trans.). Gallimard, Paris.

Marx, K. (1968). Ökonomisch-philosophische Manuskripte 1844. In *Die Heilige Familie und Schriften von Marx von Anfang 1844 bis Anfang 1845, Marx/Engels Gesamtausgabe (MEGA) III, Marx-Engels*. Verlag GmbH, Berlin.

Proclus (1965a). *Platonis Timaeum commentaria.* E. Diehl, Verlag Adolf M. Hakkert, Amsterdam.

Proclus (1965b). *Commentaire sur le Timée, Livre I.* Festugière, A.J. (trans.). Vrin/CNRS, Paris.

Proclus (1968). *Théologie platonicienne, tome 1.* Saffrey, H.D., Westerink, L.G. (eds). CUF, Paris.

Proclus (1974). *Théologie platonicienne, tome 2.* Saffrey, H.D., Westerink, L.G. (eds). CUF, Paris.

Proclus (1978). *Théologie platonicienne, tome 3.* Saffrey, H.D., Westerink, L.G. (eds). CUF, Paris.

Proclus (1987). *Théologie platonicienne, tome 5.* Saffrey, H.D., Westerink, L.G. (eds). CUF, Paris.

Sandvoss, E. (1971). *Philosophische Grundlagen der platonischen Gesetzgebung.* Musterschmidt, Göttingen.

2

From Power to Care: For an Object-Oriented Philosophy of Technology

2.1. Empirical and "thingly" turn in the philosophy of technology

A significant feature of the philosophy of technology over the last 20 years is the importance that has become attached to the concept of artifacts. Artifacts can be approached from either of two directions. On the one hand, there is a phenomenological study of artifactual mediations between humans and the world – this approach focuses on technological artifacts insofar as they actively contribute to configuring both human existence and human capacities, as well as human relations to the world (Verbeek 2005). On the other hand, there is an analysis of the components of technological artifacts, starting from the idea that they have a "dual nature" – material and intentional – and that we need to understand how the two are linked (Kroes and Meijers 2006). The "philosophy of technological artifacts" has followed both of these directions, namely post-phenomenological and analytical, simultaneously, in contrast to the so-called "classical" philosophies of technology (Brey 2010) – articulated by Heidegger, Marcuse, Jaspers, Ellul and others – that are less concerned with describing technical objects in themselves than with identifying the metaphysical and historical conditions of possibility for contemporary manifestations of technology. Through that "classical" lens, technology was seen as an expression of the desire for power, or as a reduction of all reality to that

Chapter written by Xavier GUCHET.

which is calculable and operable. The Dutch philosopher Peter-Paul Verbeek, a representative of the post-phenomenological movement, has proposed applying the Husserlian motto "back to the things themselves" to the philosophical analysis of technology, in response to the philosophies of technology of the last century, which he calls "transcendentalist", and which specifically focus on the conditions of possibility of technology to the detriment of a detailed analysis of the technological "things" themselves. Verbeek (2005) speaks of a "thingly turn" in the philosophy of technology. This emphasis on artifacts, on technological "things", contrasts with the position largely shared by "classical" philosophers, who regard technological artifacts themselves, technological "things", as philosophically uninteresting, incapable of providing a key to understanding the contemporary world. Heidegger initially set the tone: the essence of technology (that is to say, the key to understanding it) is not itself technological. French philosopher Dominique Janicaud (1987) also wrote: "[A] conceptual and reflexive approach can also be a philosophy of the 'technological object' in a sense that is philosophically minor [...]. It shows what the technological object is, but it hardly questions the techno-scientific event as such".

This recent shift in the philosophy of technology toward a focus on artifacts is part of a more general movement, dubbed by some as the "empirical turn" in the philosophy of technology. This term was coined at a workshop held in the Netherlands (in Delft) in 1998 at the initiative of two Dutch philosophers of technology, Peter Kroes and Anthonie Meijers (Kroes and Meijers 2000; Achterhuis 2001). The empirical turn in the philosophy of technology refers to a body of work by mainly American and German[1] philosophers from the late 1970s onwards. This was an attempt to move away from considering technology from an outside perspective involving either an overly prospective analysis of its consequences for the human condition and societies or an analysis of myths and fantasies surrounding it. Instead, there was a desire to link philosophical reflection to the description of technology itself. The philosopher Philip Brey provided a useful overview of this empirical turn, in which he distinguished a number of separate currents (Brey 2010). What all these currents have in common are more nuanced, less pessimistic analyses of technological development than those of the "classical" philosophers. They all also seek to challenge an

1 Mainly American and German, but not exclusively. Mario Bunge, for example was Argentinian.

assumption that is often (though not always) present in the writings of these "classical" philosophers of technology, namely that technology is autonomous and develops according to its own logic, indifferent to choices and values. Proponents of the empirical turn, on the other hand, insist on the socially constructed dimension of technology, rejecting the abstract, essentialist category of Technology with a capital T in favor of more contextual, localized approaches. The philosophy of technology in the wake of the "empirical turn" has drawn heavily on the contributions of the STS movement.

According to Verbeek, while the "empirical turn" has enabled a more detailed interpretation of contemporary technology in all its diversity and complexity, this has perhaps come at a high price: by moving closer to its object of study, and by becoming more descriptive, there is a risk that philosophy is relinquishing its critical function. Verbeek (2010) suggests that two of the trends that we may observe today are in fact responses to this risk. First, the philosophy of technology is becoming more interventionist: philosophers are now expected to provide "technology accompaniment", that is to say, to help engineers and designers make "good technological mediations". Second, ethical considerations are being given greater weight in the assessment of technology. Since technological artifacts are mixtures of materiality and intentionality, or of values, they call for an effort of description and, specifically, a critical assessment of the values and purposes that they "embed".

2.2. From technology as power to technology as care

Verbeek's analysis provides a coherent vision of the artifactual turn in the philosophy of technology and of its causes. It is possible, however, to interpret it differently, by placing it in a much longer time frame, going beyond simply a dialogue with the philosophers of the last century. If the "classical" philosophers of technology mostly neglected the objects themselves, it is because they were essentially interpreting so-called modern technology as the expression of an unbridled desire for power. The reference to an unbound Prometheus is a 20th-century leitmotiv: modern (industrial) technology is driven by the most complete *hubris* involving an uncontrolled deployment of power, and it is imperative that Prometheus be bound again through the imposition of a restrictive framework. As Hans Jonas puts it:

> The ultimately unleashed Prometheus, supplied by science with unheard-of powers and by economics with an incessant impulse, demands for ethics that, through self-restriction, prevents man's power from becoming a curse to him. (Jonas, preface to the French edition, 1990)

Jonas' position is clear: all technology, and not just modern technology, is by its very nature a manifestation of human power; indeed, it is the power peculiar to human beings. This, at least, is what we learn from the myth of Prometheus as it is most often told, starting with Plato's account in the *Protagoras* dialogue. The story is well known: in Epimetheus' division of qualities among all living creatures, man is forgotten. He is naked, completely deprived, with no power of his own (*dunamis*). To compensate for this lack of power, Prometheus gives humans the gift of fire and an ability to control it, and in so doing he gives them their own power. Technology is human power itself, but in ancient Greece this power was still weak, and in no way comparable to that of *physis* – or so says Jonas. Modern science has extended this power, making it comparable to that of nature, hence the need for a new ethics to counterbalance the power of technology. Janicaud, however, denigrates accounts of modern technology in terms of objects, because technology is, first and foremost, a manifestation of power – in the contemporary era, it is the "powers of the rational" (Janicaud 1994).

This account of the Prometheus myth remains nonetheless incomplete. From Hesiod to Plato, the myth actually tells a very different story. Here is what Hesiod says in the *Theogony*: there was a time when humans and gods feasted together, and humans needed to do nothing, not having to worry about their subsistence, since the gods provided. Grains grew by themselves, and humans could use the fire that Zeus left at the tops of ash trees. For a reason left unsaid by Hesiod, the time came when humans and gods had to be separated. Prometheus was charged with dividing an ox between gods and humans, symbolizing this separation. We know the story: Prometheus hides the edible flesh in the animal's unappetizing belly, and under a layer of shiny fat, he hides the inedible bones. Zeus has to choose: he is not fooled and sees through the deception, but he nevertheless chooses the pile of bones, which angers him. In other words, Zeus had planned to punish humankind, and he takes the opportunity provided by Prometheus to legitimize his anger and his punishment. How does he punish humankind? By depriving them of fire and grain, in other words, of their means of sustenance. In short, Zeus deprives humankind of the care the gods had previously lavished on them. Prometheus steals a spark of fire from Zeus and gives it to humankind. In his *Prometheus Bound*, Aeschylus tells a well-known story along a similar line. Prometheus is punished by Zeus for opposing his decision to eradicate the human race. No other god protested, and Prometheus alone took humanity's side. Prometheus was punished for having cared for humankind. Plato later returns to the myth of Prometheus in a dialogue not in *Protagoras*, but in *The Statesman*. While the myth of *Protagoras* introduces the theme of *techné* as human power itself, *The Statesman* tells a different story. The story is as follows: in ancient times, all living beings on Earth were under the care of divine shepherds. Every living species had its shepherd. The shepherd of humans was Zeus himself. However, the world was placed under an inexorable law, that of a cyclical reversal of its course. The world turns in one direction, then starts to turn in the other. When the rotation is reversed, the living no longer advance from childhood to old age, but from old age to childhood. The shepherds then abandon the living to their fate. The living revert to a state of nature, and chaos ensues. Zeus decides to take matters into his own hands, but in a comprehensive way, and not by reassigning a deity to care for each species. From now on, the living, including humans, must take care of themselves, and living humans are helpless. The gift of fire and the skills to master it are to enable humans to satisfy their vital needs and to protect themselves against beasts of prey – here, we find the myth in its classical version, except that *The Statesman's* narrative teaches us something decisive: *techné* is not given to humans to make up for a lack of being, an "original defect" as Bernard Stiegler puts it (Stiegler 1998), but to compensate for a lack of care. *Techné* does not rectify an ontological deficit, but an abandonment. *Techné* is what will enable human beings to take care of themselves, since their god has abandoned them. If humans are technicians, it is because they have to take care of themselves, things, living beings and their cities.

The idea of technology as power (*dunamis*) is therefore subsumed within another, more fundamental conception of technology: technology as care. In Plato, this idea of technology as care is specifically expressed through a focus on artifacts and technological things. Plato explains that technology is a creative activity. In making things, the craftsman creates something new in the world; they are the master of forms. However, if this creative activity remains unchanneled, there is nothing to stop them changing the forms of things as they see fit – they might make a table or a bed in one way today, and in another way tomorrow. The craftsman can change the forms of things, and in doing so, they risk making the world and the city very unstable – the Greeks' worst nightmare. Furthermore, for the person who has acquired it, *techné* becomes their very nature. Having addressed this problem, what does Plato say? That it is not up to the craftsman to decide the form of things, as this is fixed for all eternity by the essence of the thing in question.

> The poiema, which is the goal of his work, is beyond him... whether it was a house, shoe, flute or shield the product answered a specific natural need. It was not seen as something artificial, in the true sense of the term. Is was an eidos, a Form laid down in advance, as was the case for natural objects. The artisan did not invent it, nor could he alter it... Form is superior to the workman and to his technè, and it directs and controls the work that brings it into being. It gives this work its goals, fixes its limits, defines its framework and its means. (Vernant 2006)

Techné is knowledge of the essential form of the thing being made, and it is this knowledge that distinguishes the person who truly possesses *techné* from the person who is skilled but lacks *techné* – like the cook who, unlike the doctor, is unable to explain what they are doing or the effects they are producing. Furthermore, the possessor of a *techné* must have a well-defined place in the city, and stick to it rigorously. They will do one thing – they will be a shoemaker or a carpenter or something else, but they will do only one thing at a time, not several.

Technology is indeed a power, but this power is embedded within a requirement: the requirement of providing the human world with a stability that firmly establishes positions and roles, and with durability – conditions that enable human existence to flourish, and that are prerequisites for the existence of a public space in which free people may debate. Arendt (1958) grasped the necessity of enveloping technological power in a care for things. Labor is the manifestation of an endless, self-perpetuating power; it is a continuous process that, if we are not careful, can devour everything and ruin any form of stability in the world. The fruits of labor, on the other hand, are what channels this power, containing its destructive potential. Technological things (part of these fruits) enter a well-defined place in the world and remain there. They are what make the world solid and durable. However, for them to do their job, they need to be cared for.

2.3. Places and connections

The reason why Heidegger occupies such an important place in the philosophy of artifacts today is precisely because he described the disappearance of *techné* in the sense of care for the world – what in the past had been seen as the enveloping of human power in a complex system of relations between things, each occupying a place fixed by its nature and ensuring the solidity of world. The description of "equipment" (Zeug) in *Being and Time* (1962) has undeniable Greek traits. The equipment is "made for", used in a structure of relations that gives everything its place in a given order. It is characterized by the fact that it occupies a well-defined place in a solid, perennial structure. The tests that Heidegger associates with the disruption of the practical world cause the equipment to lose its place. When the equipment is broken, it becomes an object with no defined location: an abstract entity from the perspective of the knowing subject, itself dislocated. Another test is that of the untraceable equipment: in this case, it is no longer in its place in the original sense of the term. It ought to be there, in the toolbox, but it is not. The final test, task insistence, is also linked to a disruption of the world and a blurring of places. This picture that I need to hang troubles me, as it is out of place: it belongs on the wall, above the mantelpiece or the sideboard. The equipment is in its place, and participates in determining the places of all worldly things. The hammer is the prerequisite for the picture to be in its place – on the wall. The equipment is therefore a matter of place and order. Modern technology, on the other hand, is un-worldly, world-destroying, because it no longer gives things their rightful place. It undermines the world, because it is no longer produces things that are destined to occupy a well-defined place in the world and thus to ensure its solidity and durability. Like the equipment, modern technology can also be interpreted as a series of "transmissions" – the flow of the Rhine is "transmitted" to the turbines of the power station, which "transmit" it to the electricity distribution network, which in their turn "transmit" it to the industries and homes it supplies with energy; but this structure of transmissions is no longer based on assigning each thing a place determined by its nature and purpose. The transmissions characteristic of modern technology mean the disruption of place, giving way to an all-pervasive placelessness (Heidegger 1977). The little stone bridge defines and creates a place; the power station is the absolute non-place. No place is created by it. It is no longer even the object. The placelessness of the power station is not of the same nature as the placelessness of the broken hammer: for the hammer, placelessness is a challenge that can reveal the hammer's essence. The placelessness of the hammer is merely the expectation of a return to its place – the hammer can be repaired or replaced. The placelessness of the power plant is something else: it is the loss of the very possibility of place. Technology no longer provides the foundation for the world. It is the manifestation of human power, the destroyer of the world. Extracting, storing, transporting and supplying power: this is the overall dynamic of modern technology. Everything is caught in the vice of the *Gestell*, directed toward the provision of power. Any being is standing in reserve, in other words, an absolute non-place: the destruction of every possible place, swallowed up in the cycle of production and consumption sustained by power.

The philosophy of artifactuality, whether it is approached from a post-phenomenological or from an analytical direction, reflects a desire to give a more empirical orientation to a philosophical reflection on technology. At a deeper level, however, it revives the ancient Greek idea of *techné* as caring for the world – except that this care for the world, importantly, is no longer ordered by notions of stability and fixedness of places and roles. Our technical objects are characterized less by the fact that they occupy a well-defined place than by their trajectories and the connections that they forge with all kinds of entities, animate or inanimate, at different scales of space and time. Thus, conceptualizations of the technological artifact beyond even post-phenomenological and analytic approaches to artifactuality (such as Michel Serres' concept of quasi-objects (1997); Bruno Latour's concept of the technical object as a "fold" (2002); Gilbert Simondon's concept of technical objects (2017); Timothy Morton's concept of hyper-objects, (2013)) all lean toward thinking of technical "things" less in terms of their defined place in an ordered world but instead in terms of their trajectories in an ever-changing world.

Taking care of the world means keeping a watchful eye on everything to which our technological artifacts are linked. In this respect, an object-oriented philosophy of technology is a step toward thinking about the environmental sustainability of our technologies.

2.4. References

Achterhuis, H. (ed.) (2001). *American Philosophy of Technology: The Empirical Turn*. Indiana University Press, Bloomington.

Arendt, H. (1958). *The Human Condition*. The University of Chicago Press, Chicago.

Brey, P. (2010). Philosophy of technology after the empirical turn. *Techné*, 14(1), 36–48.

Heidegger, M. (1962). *Being and Time*. Blackwell Publishers Ltd, Oxford.

Heidegger M. (1977). *The Question Concerning Technology and Other Essays*. Garland Publishing Inc., New York and London.

Janicaud, D. (1987). Des techniques à la techno-science : l'enjeu philosophique. *Revue Internationale de Philosophie*, 161(2), 184–196.

Janicaud, D. (1994). *Powers of the Rational. Science, Technology, and the Future of Thought*. Indiana University Press, Bloomington.

Jonas, H. (1990). *Le principe responsabilité : une éthique pour la civilisation technologique*. Flammarion, Paris.

Kroes, P. and Meijers, A. (eds) (2000). *The Empirical Turn in the Philosophy of Technology*. Elsevier Science Ltd, Oxford.

Kroes, P. and Meijers, A. (2006). The dual nature of technical artifacts. *Stud. Hist. Phil. Sci.*, 37(1), 1–4.

Latour, B. (2002). Morality and technology. The end of the means. *Theory, Culture and Society*, 19(5/6), 247–260.

Morton, T. (2013). *Hyperobjects. Philosophy and Ecology after the End of the World*. University of Minnesota Press, Minneapolis.

Serres, M. (1997). *Genesis*. University of Michigan Press, Ann Harbor.

Simondon, G. (2017). *On the Mode of Existence of Technical Objects*. University of Minnesota Press, Minneapolis.

Stiegler, B. (1998). *Technics and Time, 1. The Fault of Epimetheus*. Stanford University Press, Stanford, California.

Verbeek, P.-P. (2005). *What Things Do. Philosophical Reflections on Technology, Agency and Design*. Pennsylvania State University Press, University Park.

Verbeek, P.-P. (2010). Accompanying technology: Philosophy of technology after the ethical turn. *Techné: Research in Philosophy and Technology*, 14(1), 49–54.

Vernant, J.-P. (2006). *Myth and Thought among the Greeks*. Zone Books, New York.

3

Thinking in the Anthropocene Era with Henri Bergson

In this chapter, we start from a premise: the Anthropocene is not separate from a certain philosophical trend that began with Plato and is still alive today. This trend is set around a vertical ontology that places humans at the center of the world and at the top of the scale of existing beings, attributing technical capacities to us that exclude all other living beings, which we now know can structurally transform the environment and, beyond that, the various systems that regulate the Earth as a whole.

Descartes' famous statement that this technical (and scientific) capacity – which has now become a power – has made "man like the master and possessor of nature" (Descartes 1637), legitimizing humanity's hold over nature, whether in terms of knowing or acting, is the first account of this "age of humans", the effects of which have only been felt for half a century. There is therefore a profound correlation between this philosophical trend and what we call the "Anthropocene", from which it would seemingly take too long to escape. Nevertheless, if we are geologically imprisoned today, a deconstruction of this vertical ontology could perhaps enable us to take a psychological, political, economical and philosophical step out of the Anthropocene. This follows the proposal of Sébastien Dutreuil, who suggests that the Anthropocene must first be understood as a political, philosophical and pragmatic concept rather than a geological and epistemological one (Dutreuil 2018).

Chapter written by Élise LAMY-RESTED.

The first step is to reintegrate technology into the world of the living, by showing how it is an extension of it, as humanity has monopolized its power to ensure its domination. From this point of view, *L'évolution créatrice* (Bergson 1907) is the first philosophical work written in a context of overpowering technology that puts us on this path, despite Begson's creation of the *homo faber* concept, which distinguishes humans, who are the manufacturers of technology, from other living beings. In order to understand how Bergson deconstructs *homo faber*, we need to explore each layer of his philosophy, which, while connected to one another, nevertheless benefit from a form of independence. The confinement of reality into categories certainly has an explanatory or practical effectiveness on which the philosophy, science and technology of *homo faber* could be based, but it remains far removed from the creative power of the *élan vital* (vital impulse) which, according to Bergson, is the principle of all existing things, whether they are organic or inorganic. Bergson's philosophy thus requires a speleological reading, progressively taking us into the depths of a life that is very different from that of *homo faber*, with its intelligent but superficial life, full of knowledge but devoid of intuition. The life of *homo faber* is pretentiously poor and far removed from the vital, creative and unorganized impulse.

3.1. *Homo faber*

> If we could strip ourselves of all pride, if, to define our species, we strictly adhered to what history and prehistory present to us as the constant nature of man and intelligence, we would perhaps not say *homo sapiens*, but *homo faber*. When all is said and done, *intelligence, seen in its original light, is the ability to make artificial objects, especially tools, and to vary their manufacture indefinitely.*[1] (Bergson 1907, p. 140)

1 This is a personal translation of the following original quote: "Si nous pouvions nous dépouiller de tout orgueil, si, pour définir notre espèce, nous nous en tenions strictement à ce que l'histoire et la préhistoire nous présentent comme la caractéristique constante de l'homme et de l'intelligence, nous ne dirions peut-être pas *homo sapiens*, mais *homo faber*. En définitive, *l'intelligence envisagée dans ce qui en paraît être la démar-che originelle est la faculté de fabriquer des objets artificiels, en particu-lier des outils, et d'en varier indéfiniment la fabrication*" (Bergson 1907, p. 140).

While reintroducing humanity into the world of the living, *L'évolution créatrice* makes humanity a being in its own right, whose intelligence enables the organization of matter to make tools. In so doing, it reverses an entire philosophical tradition that identifies intelligence or understanding with an ability of knowing or thinking that is a priori in Kant's terms. According to Bergson, "intelligence is first and foremost about manufacturing"[2] (Bergson 1907, p. 154). While instinct innately enables animals to be in tune with their external environment, intelligence connects humans to their environment through the mediation of consciousness and representation. In animals, action is thus confused with the instinctive knowledge of how to behave, as well as matter in their environment, which they modify or adapt to without being aware of it. Humans' survival depends on their ability to visualize forms, which precedes action: the decomposition or recomposition of matter according to the form imagined enables the transformation of the environment in their own image. Intelligence therefore directly deals with inert matter, to which it can give a plethora of forms independently of its initial shape. The very essence of intelligence is to treat everything, whether living or non-living, as matter over which it can have total control, and to which it can give every possible form of organization. In other words, intelligence is fixist: destined to organize inorganic matter, it will treat living things in the same way, denying their becoming and reifying them. From such a perspective, it would appear that only man, who is also the only living being endowed with intelligence, is a technological being or a *homo faber*.

3.2. Intelligence as an instinct

However, even if intelligence belongs to humans alone, it is indeed a product of the *élan vital* (vital impulse), not an ability that "fell from the sky with its form" (Bergson 1907, pp. 152–153):

> All the elementary forces of intelligence tend to transform matter into an instrument of action, in other words, in the etymological sense of the word, into an *organ*. Life, not content with producing organisms, would like to give them the appendage of inorganic matter itself, converted into an immense

2 Original quote: "L'intelligence vise d'abord à fabriquer" (Bergson 1907, p. 154).

organ by the industry of the living being. This is the task it initially assigns to intelligence.[3] (Bergson 1907, p. 162)

Intelligence is therefore not an ability that is independent of life, reduced to a set of mechanical processes. It is a force for making and acting, whose aim is not to change matter, whether it be inert or living, into a form that can potentially move mechanically, but rather to manufacture something organic out of the inorganic, in other words, to make life. When we previously mentioned that intelligence was fixist, we were only considering the most superficial layer of intelligence. In actual fact, intelligence is dynamic and inventive, not in the sense that it can be converted into a multiplicity of forms or representations that give rise to the diversification of technical objects, but in the sense that its essential tendency is to prolong life in inorganic matter. In other words, intelligence is itself an *organ*, that is to say, an instrument of action produced by life, which thus gives itself the capacity to externalize itself.

Just as life has developed endosomatic technical instruments in animals, such as the crab's claw, as Leroi-Gourhan illustrates, life has also given itself the ability to manufacture exosomatic instruments, in other words, removable tools or external machines, by developing the organ of intelligence. Intelligence is thus the endosomatic organ that enables us to manufacture exosomatic and organic objects, despite their initially inert and inorganized nature. Although Bergson asserts that only humanity is endowed with an intelligence capable of mechanical invention, the fact remains that intelligence is not the queen faculty that essentially distinguishes humans from other living beings. In the same way that the crab's claw is an organ, intelligence is an organ whose specificity lies in its ability to produce artificial objects. In Bergson's (1907) words: "Accomplished instinct is the ability to use and even construct organized instruments; accomplished intelligence is the ability to make and use inorganized instruments"[4] (p. 141).

3 This is a personal translation of the following original quote: "Toutes les forces élémentaires de l'intelligence tendent à transformer la matière en instrument d'action, c'est-à-dire, au sens étymologique du mot, en *organe*. La vie, non contente de produire des organismes, voudrait leur donner comme appendice la matière inorganique elle-même, conver-tie en un immense organe par l'industrie de l'être vivant. Telle est la tâche qu'elle assigne d'abord à l'intelligence" (Bergson 1907, p. 162).

4 This is a personal translation of the following original quote: "L'instinct achevé est une faculté d'utiliser et même de construire des instruments organisés ; l'intelligence achevée est la faculté de fabriquer et d'employer des instruments inorganisés" (Bergson 1907, p. 141).

Instinct and intelligence thus pursue the same objective: to enable living beings to evolve in their environment by modifying it. What remains to be understood is in what sense they can be defined as "two forms of psychic activity". According to Bergson (1907), this common dynamic is that of "Consciousness in general", which he writes with a capital letter (p. 187), and which must be redefined here. Indeed, as we descend a little further into the depths of life, we no longer perceive Consciousness as an instrument used to manufacture artificial objects by representing the forms to be given them, but as an organizing power whose access to representation is the most accomplished and remote possibility of this power. In Bergson's (1907) words: "It is as if a broad current of Consciousness had penetrated matter, charged, like all Consciousness, with an enormous multiplicity of virtualities that were intertwined" (p. 182). This Consciousness in general is therefore part of a time frame that is not that of consciousness understood as intelligence. These virtualities constitute all possible tendencies, not all of which have undoubtedly been realized, but all of which tend toward organization, regardless of form, instinct or intelligence. This enveloping Consciousness never ceases to work on the tendencies that realize it, each of these tendencies realizing it according to its own particular genius. Thus, according to Bergson, we find forms of consciousness in animal and plant instinct, while human intelligence also has its instinctive sides.

As an organ produced by life, intelligence innately possesses the means to manufacture, in other words, the means to extricate itself from this innateness. This means that the instinct inherent in consciousness or intelligence would be the tendency to stop acting on instinct. The same applies to instinct, which possesses traits of intelligence or consciousness. Acknowledging that "we do not hesitate to declare certain animals intelligent"[5] (Bergson 1907, p. 138), particularly those that manage to fashion an instrument or use human-made technical objects to their advantage, Bergson understands this common sense in the development of his philosophy. If Consciousness in general is the momentum that is at the origin of divergent tendencies such as instinct and intelligence, then it would make sense that both animals and plants virtually possess this Consciousness, whose full actualization is intelligence. Consciousness either lies dormant in instinct or tends toward its actualization

5 Original quote: "Nous n'hésitons pas à déclarer intelligents" (Bergson 1907, p. 138).

by prompting the animal to act in an "intelligent" way. When an animal uses a human-made tool appropriately, or distinguishes an artificial object from a natural one, it is demonstrating intelligence and a natural born capacity for invention. As Bergson (1907) says, "invention becomes complete when it materializes in a manufactured instrument. This is where animal intelligence tends towards an ideal"[6] (p. 139).

The unified tendency that produces both instinct and intelligence is organization. While inert matter is, by definition, unorganized and inorganic, living matter is distinguished through its efforts to organize. Where instinct organizes matter in a non-conscious way, intelligence does so consciously, passing through the filter of representation to invent and give new form to an object. According to Bergson (1907), whom we quote at some length:

> Intelligence is characterized by a natural lack of understanding of life. Instinct, on the other hand, is molded on the very form of life. While intelligence processes everything mechanically, instinct proceeds organically, so to speak. If the Consciousness that slumbers within it were to awaken, if it were to internalize knowledge instead of externalizing it into action, if we knew how to question it and if it were able to answer, it would reveal the innermost secrets of life. It merely continues the work by which life organizes matter, to such an extent that we cannot say, as has been shown so often, where organization ends and instinct begins.[7] (p. 166)

---

6 This is a personal translation of the following original quote: "L'invention dev[enant] complète quand elle se matérialise en un instrument fabriqué. C'est là que tend l'intelligence des animaux comme à un ideal" (Bergson 1907, p. 139).

7 This is a personal translation of the following original quote: "L'intelligence est caractérisée par une incompréhension naturelle de la vie. C'est sur la forme même de la vie, au contraire, qu'est moulé l'instinct. Tandis que l'intelligence traite toutes choses mécaniquement, l'instinct procède, si l'on peut parler ainsi, organiquement. Si la conscience qui sommeillait en lui se réveillait, s'il intériorisait en connaissance au lieu de s'extérioriser en action, si nous savions l'interroger et s'il pouvait répondre, il nous livrerait les secrets les plus intimes de la vie. Car il ne fait que continuer le travail par lequel la vie organise la matière, à tel point que nous ne saurions dire, comme on l'a montré bien souvent, où l'organisation finit et où l'instinct commence" (Bergson 1907, p. 166).

3.3. Life as an organization

Organization is thus the basis of life's activity, even before its individuation. Indeed, life tends to "individuate" in Simondon's (1964) words, that is to say, to become autonomous in a differentiated organism that is never closed in on itself: "Individuality comprises an infinite number of degrees and [...] nowhere, not even in man, is it fully realized"; we must "see this as a characteristic property of life"[8] (Bergson 1907, p. 12).

If the individual can be considered as an organism, in other words, a whole capable of acting in and with its environment, modifying or incorporating it, whether by instinct or intelligence, organization is the tendency that produces individuals. The nature of organization, which is found in all living beings, whether or not they are individuated, is duration or becoming. It is a tendency that projects itself into the future, with no other end in view than to persist in its being. In other words, as a tendency, life enjoys a continuous temporality: the very nature of a tendency is to be simultaneously past, present and future. It is future, because it is a projection of life that wishes to endure. As a result of it being a projection, this unrepresentable objective is entirely understood in the past and in the present. It is understood in the present in the sense that it simultaneously exceeds and organizes it. It is understood in the past in the sense that all the virtualities that constitute it tend toward organization.

For Bergson (1907), temporality is thought of in terms of the past, which is why he identifies lasting life with memory: "Duration is the continuous progress of the past, which gnaws at the future and swells as it advances" (p. 4). Memory is not a collection of memories, but a continuous transformation of the present as a result of an ever-living past that looks to the future. The past is, in fact, a vessel of virtualities that are only partially actualized in a ramification of different tendencies such as instinct and intelligence. However, the fact that these virtualities are not actualized does not mean that they have been erased: they are preserved pending their possible actualization and are transformed in accordance with the present context. By continuing to work on the present, these virtualities prevent it from closing in on itself. In other words, according to Bergson, the past

8 Original French version: "[...] l'individualité comporte une infinité de degrés et que nulle part, pas même chez l'homme, elle n'est réalisée pleinement. Mais ce n'est pas une raison pour refuser d'y voir une propriété caractéristique de la vie" (Bergson 1907, p. 12).

organizes the present and the future. The present is modified by the past, which has never disappeared, and the future is shaped by the work of the past in the present. Thus, "the past preserves itself automatically. All of it, no doubt, follows us at every moment: what we have felt, thought, wanted since our earliest childhood is there, leaning on the present which is about to join it, pressing against the door of consciousness that wants to keep it out"[9] (Bergson 1907, p. 5). Although Bergson seems to be talking about human subjectivity here, it would be a mistake to assume that memory is only limited to beings endowed with consciousness, in the strict sense of the term. Bergson maintains that life itself is memory. Thus: "The evolution of the living being, like that of the embryo, implies a continuous recording of time, a past that persists in the present, and consequently an appearance at least of organic memory"[10] (Bergson 1907, p. 19). The organization, organism or organ is thus defined by its temporality, which is that of memory or duration, in other words, a temporality of the past.

Put another way, the past is an organization that must be immediately distinguished from order or planning. While both order and planning are determined, that is to say, capable of being represented, organization is an internal arrangement whose sole purpose is to bring living beings into contact with their external environment. In other words, organization is technical in the sense that it is an instrument of action that enables living beings to modify and incorporate their external environment. In Bergson's (1907) words:

> It is up to life to create a form for itself that is appropriate for the conditions it faces. It will have to take advantage of these conditions, neutralize its disadvantages and use its advantages, as well as respond to external actions by building a machine that bears no resemblance to the latter. In this case, adapting

9 This is a personal translation of the following original quote: "Le passé se conserve de lui-même, automatiquement. Tout entier, sans doute, il nous suit à tout instant : ce que nous avons senti, pensé, voulu depuis notre première enfance est là, penché sur le présent qui va s'y joindre, pressant contre la porte de la conscience qui voudrait le laisser dehors" (Bergson 1907, p. 5).

10 This is a personal translation of the following original quote: "L'évolution de l'être vivant, comme celle de l'embryon, implique un enregistrement continuel de la durée, une persistance du passé dans le présent, et par conséquent une apparence au moins de mémoire organique" (Bergson 1907, p. 19).

> no longer means repeating, but replicating, which is quite different.[11] (p. 58)

To replicate is to respond to a situation in an organized, instinctive or intelligent way. Regardless of the modality of this response, Bergson does not think of it in terms of a mechanical reaction, even if this response paradoxically calls for the "construction of a machine". If we go down to the deepest layer of life, we understand that if humanity is *homo faber*, it is because life is the technical tendency that makes machines to survive. However, these machines differ from human-made machines in that they are not the result of the organization of inorganized matter.

If intelligence builds mechanical technical objects that reveal its relationship with inert matter and its inability to produce life, life builds organic machines. In other words, the machines built by life are organs that build other organs or machines that respond to the demands of the external environment, not mechanically, but organically. It is in this sense, and at this level of thought, that Canguilhem (1952) rightly asserts that: "*L'évolution créatrice* is […] a treatise on general organology" (p. 125).

3.4. Conclusion: the power and limits of general organology

General organology, defined as an extension of life that invents organs in the etymological sense of "instruments" to adapt to or interact with the external environment, represents a first deconstruction of this vertical ontology, which Stiegler reintroduces by identifying general organology with the process of individuation that is specific to the human being, who, in the strict sense and in their own right, is a technical being (Stiegler 1994a). It allows us to think of the emergence and technological power at work in every organism that constructs itself in relation to other existing beings, whether living or non-living. By inventing this concept of "general organology" from his reading of *L'évolution créatrice*, among other texts, and by granting all living beings a certain technical capacity and implicitly

---

11 This is a personal translation of the following original quote: "C'est à la vie qu'il appartiendra de se créer à elle-même une forme ap-propriée aux conditions qui lui sont faites. Il va falloir qu'elle tire parti de ces conditions, qu'elle en neutralise les inconvénients et qu'elle en uti-lise les avantages, enfin qu'elle réponde aux actions extérieures par la construction d'une machine qui n'a aucune ressemblance avec elles. S'adapter ne consistera plus ici à répéter, mais à répliquer, ce qui est tout différent" (Bergson 1907, p. 58).

proposing to reintroduce our contemporary technology within the limits of a shared reality, Canguilhem gave us a decisive incentive to think outside the Anthropocene.

Élan vital (vital impulse) cannot be reduced to a general organology that identifies life with an organization, essentially distinguishing it from so-called inert matter. As we have tried to show elsewhere by introducing the notion of "technical life" in a new way (Lamy-Rested 2023), this notion – before being re-embraced by Stieglerian philosophy – deconstructs the vertical ontology that has long influenced a major trend of philosophy, giving us the means to think of a horizontal ontology, or perhaps, more precisely, a cosmology that is essentially another way of interacting with other existing beings, whether they are "living" or "non-living".

3.5. References

Bergson, H. (1907 [2007]). *L'évolution créatrice*. PUF, Paris.

Canguilhem, G. (1952). *La connaissance de la vie*. Vrin, Paris.

Descartes, R. (1637 [2016]). *Discours de la méthode*. Garnier-Flammarion, Paris.

Dutreuil, S. (2018). L'anthropocène est-il un concept d'histoire de la Terre ? Le nom qui ne dit pas son épistémologie. In *Penser l'Anthropocène*, Beau, R., Larrère, C. (eds). Presses de Sciences Po, Paris [Online]. Available at: https://hal.archives-ouvertes.fr/hal-01789716/document.

Lamy-Rested, E. (2023). La vie technique est-elle une organologie générale ? In *Objets vivants. Aux frontières de la vie*. Éditions Mimésis.

Simondon, G. (1964). *L'individuation à la lumière des notions de forme et d'information*. Millon, Grenoble.

Stiegler, B. (1994a [2018]). *La technique et le temps, T.1*. Fayard, Paris.

Stiegler, B. (1994b [2018]). *La technique et le temps, T.2*. Fayard, Paris.

PART 2

Epistemological Challenges of Modern Technologies

Introduction to Part 2

This second part involves technology's scientific aspect. Today, our technical tools have acquired unprecedented power and granularity. Since the 1970s, when personal computers first came about in the world of work, their use has profoundly transformed our forms of life – communication and payment capabilities, access to knowledge and entertainment – and all facets of existence – professional, economic, social or intimate – have been affected at an unrivalled pace. By comparison, the spread of the steam engine looks like the tortoise struggling against Achilles' stride, without the support of Zeno of Elea's illusory paradox.

In the face of these changes, there are critical epistemological questions. What do contemporary technologies produce, and how solid, fragile, flawed, misunderstood or even misconceived are they? The speed of innovation and the complexity of the tools used mean that very few scholars are in a position to judge the epistemological robustness of the solutions adopted. The problem is therefore not only theoretical, but also practical, and is proving decisive insofar as the use of our technical capabilities can modify our future in both time and form. Our capacity for insight is essential if we are to tackle this issue, and distinguish between what are changes, or even crises, and what are merely transient or superficial states.

The end of scientific theory[1] has been announced, and thanks to our technical tools, scientists have never had such a capacity for observation and

---

Introduction written by Valérie CHAROLLES.

1 Anderson, C. (2008). The end of theory: The data deluge makes the scientific method obsolete. *Wired*.

such a level of effective observation. Neither Leonardo da Vinci nor René Descartes imagined that reality could be squared to such an extent. However, in order for data scientists to be able to proceed in a tried and tested way, they need to be assured of a sufficiently large and unbiased corpus of observations, which is rarely the case.

The advent of "artificial intelligence" has also been announced. This term, proposed in 1955 and adopted by the computer science community in 1956[2], originates from Alan Turing's 1950 article "Computing Machinery and Intelligence"[3]. However, as early as the 1980s, John Searle demonstrated the poorness of this intelligence, which did not deserve the name, using the Chinese room experiment[4]: what is an exchange without understanding or intention? Nothing resembling intelligence.

We will not be able to clarify all the twists and turns that technologies reserve for the solidity of proven knowledge here – be it through falsification, verification or renewed experience. We will focus on a few distinctive elements: how are technology, knowledge and actions linked? To what extent does technology call our models of knowledge into question today? These are the questions that we will provide answers for, without losing sight of the fact that epistemological issues run throughout this book, and are to be found in the metaphysical, ontological and political problems that form the framework of the other sections.

Bruno Bachimont opens the debate with "The Code Paradigm: Trace Amnesia and Arbitrary Interpretation". In this chapter, he shows how digital technologies and their computer coding constitute a new age of knowledge. After knowledge by similarity, then knowledge by causality, the code paradigm shows a disruption, a new form of relationship between what knows and what is known: by offering a universal way of expressing knowledge (everything is codable) and manipulating it (everything is calculation), it does not simply provide a universal science, but, through the diversity of codes and calculations, a universal means of describing the world. However, this description is not an explanation, unless it is linked to a model, and requires a narrative and storytelling in order to make sense of it.

2 John McCarty invented the name; 1956 lecture at the University of Dartmouth, for its prosperity.

3 Turing, A.M. (1950). Computing machinery and intelligence. *Mind*, 49, 433–460.

4 Searle, J. (1984). *Minds, Brains and Science*. Harvard University Press, Cambridge.

This paradigm of knowledge thus leads to a curious coupling between calculation and storytelling; this paradigm only adheres if calculation and storytelling lead to the same picture.

In "“Motion” Machines and “Token” Machines: Milestones in the History of the Alphabet", Jean Lassègue continues the analysis of the epistemological moment and challenges the misleading idea of “artificial” intelligence. Relying in particular on François Sigaud, he shows us that the deployment of “token” or “motion” machines is not a natural and individual movement, but a process linked to social and collective organizations. He distinguishes three stages in the history of “token” machines (abjad, alphabet, code); the last, that of computing, highlights the automation of writing and the empowerment of the machine in the production of tokens, as opposed to “motion” machines, whose operation depends on their power supply. This difference between “token” and “motion” machines prevents misunderstandings about the abilities of the computer, just as it prevents us from attributing the abilities of token machines to motion machines, particularly autonomy.

In "“Digital Technology”, Revealing Intersections between Epistemology, Political Philosophy and Philosophy of Technology", Éric Guichard finally explores five theses linking different areas of philosophy. Like Jean Lassègue, he treats writing as a technique whose mastery is essential. He defines the Internet and digital technologies as a transformation of writing. He then shows how our thinking has always been essentially technical and highlights the epistemological-political crisis of the present moment: the abduction of writing by big digital companies. He uses political philosophy to take on the role played by these actors, a role that the triptych of states, institutions and citizens does not capture.

4

The Code Paradigm: Trace Amnesia and Arbitrary Interpretation

4.1. Introduction

While it is commonplace to note that our lives, whether material, social, intellectual or cultural, are conditioned by the use of digital technology and instruments, it is often more difficult to determine the scope and depth of the changes or evolutions we observe. In this chapter of the book, we focus on the nature of knowledge and the relationship we more or less implicitly establish when we assert *this* knowledge about *that*: what relationship do we then understand between *this*, which is the knowledge, and *that*, which is what is known? More specifically, how do they differ from one another, and how does the evolution of what is known translate into a joint transformation of the knowledge that takes it as its object?

What we want to explore here is how digital technology introduces a new paradigm of knowledge, in other words, a new type of relationship between knowledge and the object of this knowledge (what is known). This relationship is code. Of course, we did not wait until the digital era to mobilize codes, establishing them both to encode and to decode content, but it had rarely crossed the minds of their designers that the purpose of code was to establish an epistemic link with what is coded. On the contrary, most of the time, code is what is used to make the link between knowledge and its object obscure and inaccessible: code has to be broken, or, at the very least,

Chapter written by Bruno BACHIMONT.

analyzed, in order to find a formulation that can establish the link to be known, this link no longer being of a coded nature.

There is therefore a kind of paradox in our contemporary use of digital technology and its massive presence in the intellectual techniques of writing and naming, such as in Lévy's (1990) sense, implemented in the diversity of operations of the mind. If what makes it possible to hide knowledge has been placed at the service of what makes it possible to express knowledge and convey its gnoseological content (what it carries and what makes it indicate properties that are truly attributed – in the sense of being controllable and verifiable – to its object), then there must have been a radical gain enabling such a shift. We argue that this gain is the arbitrariness of the sign and code, which makes it possible to envisage a universal way of expressing knowledge (everything is codable) and a universal way of manipulating it (everything is calculation). By adopting a coded conception of knowledge – knowledge being a code for reality – our contemporary era proposes a universal approach to the world, homogenizing its expression and transformation via code and calculation. Therefore, instead of having a single, universal science, we have a single, universal way of describing the world through the diversity of codes and calculations.

To locate this code paradigm in its genesis and its consequences, we first look at the other paradigms that can be encountered to formulate knowledge and the known so as to explain the relationship between digital technology and code and ultimately conclude with the consequences that this involves when we want to mobilize coded material as an expression of knowledge. We then argue that the appropriation of knowledge requires the mobilization of tools other than code, particularly a narrative, so that the contemporary code paradigm is a coupling between calculation and narrative. This coupling is bound to raise questions, which we address in the conclusion section.

4.2. The ages of knowledge

As soon as knowledge is created, it is manifested through an expression that relates to an object of knowledge. This introduces a particular characterization of knowledge, in which the concept is not aimed at the ability to do something (in the sense of being able to perform an action, solve a problem, respond to a situation), but at the expression or representation of this ability through a given mode of expression.

From this perspective, we can consider that our relationship with knowledge has gone through three major successive periods. According to the first, to know is to have a representation that resembles the known object: the gnoseological link consists of a resemblance between knowledge and the object of knowledge. It only contains truth about the object in proportion to its resemblance to it. This age is based on the ancient and medieval ages and can be broken down into approximate periods.

According to a second age, knowledge consists of a causal link. The gnoseological link between knowledge and its object is that of causal production. Knowledge no longer resembles its object, but must have a necessary link to it as its origin or source. This is the classical age of knowledge, and to some extent, still ours.

Lastly, the third age is that of coding, where knowledge is an arbitrary, manipulable code of the known object. Neither in the image of the latter nor in a causal relation with it, knowledge is manipulable codification whose link with the object is a practice of coding and decoding. Let us specify this.

4.2.1. *The age of resemblance*

Although antiquity had more than one theory of knowledge and potential associated paradigms, the Latin Middle Ages chose the theory of resemblance (Hamou 2002), thus ensuring continuity with what antiquity had been able to suggest. According to this approach, knowledge is a form shared between the known object and the knowing subject, with the abstract form of the former (in the literal sense, detached from it) informing the mind of the latter, who then possesses a simulacrum (visual representation or "species") resembling the known object, given that it shares the same form and therefore the same appearance (Pasnau 1997). While the matter of the object is obviously distinct from that of the subject, these two poles of knowledge come together in a common form generated by a common act: the act of the perceiver and the perceived, the knowing and the known.

Such an approach leads to a very specific question and research agenda. For example, while we can understand in the case of perception that the form of an object is imprinted on our senses, this becomes more difficult if we consider intellection rather than perception. What does it mean when the mind receives the form of the known object? What would be the intelligible species playing the role of the sensible species in the case of conceptual knowledge? Also, how do we answer the objection that, in this approach, the mind is content to know its representations, even if they are common to objects, rather than the objects themselves?

We do not intend to go any further into these considerations, which can become highly technical. We are satisfied that this approach enables us to establish a type of gnoseological link between knowledge and its object, and the posture that derives from it with regard to understanding and thinking: in such a paradigm, if to know is to establish a resemblance, then to understand is an act of unveiling (we show what is in the image of things) and to think is to contemplate.

4.2.2. *The age of causality*

While the resemblance paradigm has been debated from the outset, its abandonment and the adoption of the causality paradigm was the work of the classical era, particularly of Descartes (Guenancia 2000). Indeed, Descartes' criticism of the ancient and medieval paradigm was that resemblance itself bore no gnoseological relation. Having the same image makes no more sense than having the same name: by abandoning the idea that having the same image should suggest that the objects under consideration must possess the same essence, and therefore the same ontological explanation, it was necessary to propose another way of combining knowledge and its object.

At the foundation of what was to become modern scientific knowledge, the new paradigm rested on the principle that knowledge is knowledge if it is the causal consequence of its object, thus in turn becoming its manifestation and expression. By demonstrating the causal relationship, for example, between the object seen and what appears in the mind, Descartes clearly shows that the result cannot resemble its origin, but has properties whose systematicity reproduces that of the perceived object. Physical and perceptual resemblance is now contrasted with isomorphism, in the mathematical sense of properties that behave analogously between the knowing and the known, without them having to share any significance resemblance.

From this perspective, to understand is to reproduce: we have understood a phenomenon when we are able to reproduce it experimentally. To think about the phenomenon is to explain it, in other words, to unfold the causal chain of events leading from the object as cause to knowledge as consequence.

4.2.3. *The age of coding*

While modern science has largely been built on the notion of causality and representation caused by what is represented, and not in the image of the latter (with variations and nuances that Daston and Galison (2012) detail perfectly), the status of representations has evolved when it came to thinking about their manipulability; in other words, the ease of calculating with and from them. Indeed, representations with a scientific value are part of the experimental and computational paradigm launched by Galilean–Cartesian science. Measurements corresponding to initial data enable a calculation that predicts the measurements associated with the resulting phenomena.

For the actual implementation of science and technology, the need for calculation soon became very significant (Gingras et al. 1998): whether it was tables giving the values of the main mathematical functions (logarithm tables, for example) or physical measurements (data on materials, astronomical measurements, etc.), whole workshops of calculators were mass-producing calculated results. In order to avoid errors, it was essential to break down calculations in the most elementary way, so that each step could be carried out with the least possible risk of error. Following the example of Pascal and Leibniz's calculating machines (Pratt 1995; Wagner 1998), the aim was to automate calculations, even if they ultimately had to be carried out by human beings.

When theories of computation emerged in the 1930s (Copeland et al. 2015), particularly the one proposed by Turing through the notion of the ideal machine, now known as the Turing machine, we formally and mathematically possessed what practice had already implemented for its encrypted data needs. From then on, calculation became the knowledge paradigm: knowledge is recognized as such when it can be calculated, and therefore it is the result of prior coding.

4.3. Digital technology and coding

What these new computability paradigms introduce is, among other things, a new design of character signs. Indeed, if we consider the formalism of the Turing machine, for example, the latter consists of the manipulation of symbols defined according to a finite or countable alphabet. In particular, these symbols are characterized by two essential properties: on the one hand, their type and, on the other, their distinguishability. According to the first, as soon as we encounter a symbol, we know what type it is: this "a", for example, has the same type as that other "a", namely, the letter /a/. According to the second, the only thing that defines type /a/ is that it is distinguishable from other types, for example type /b/ and type /c/ if we consider a three-symbol alphabet. Of course, the preferred alphabet in digital technology is binary and made up of two types of symbols commonly referred to as 0 and 1 (but we could have called them "Tic" and "Tac", or "Bonnie" and "Clyde", while respecting the aforementioned constraints on typing and distinguishability).

These two properties of typing and distinguishability have significant consequences in terms of our interests here: there is no case where a character sign manipulated by a computing machine is characterized by what it signifies; a character sign without meaning, a signifier that means nothing, as it were. This is why ordinary language is so misleading, because it could be mistakenly assumed that calculation is about representations. However, this is not the case: symbols manipulated during calculation represent nothing in themselves. Similarly, the calculation rules used are equally meaningless; they merely indicate how, from a certain number of symbols, another symbol is to be deduced, the latter then being understood as, for example, the addition or multiplication of the preceding ones.

In this way, we understand that digital technology, in the sense of a calculation performed by a Turing machine, is reduced to a blind manipulation of symbols emptied of their value. This is essential if the manipulation is to be automated and assigned to a machine. However, the meaning of symbols that are ignored when used as substrates for a calculation that manipulates them must be considered, if the results of this calculation are to be exploited. We therefore need an agreement for reading the symbols we manipulate: the importance here is that the agreement is necessarily external, arbitrary and indifferent to the nature of the symbols under consideration. So much so, in fact, that the same calculation can be interpreted in different ways, depending on the agreement: this is how popular wisdom taught us that we should not mix cabbages and carrots in our calculations, precisely because there is nothing in a calculation that forbids us from doing so. It is a matter of agreement. On a more technical note, we can see that a digital bitstream can be read by both an audio player (we will hear something) and a video player (we will see something). Binary content does not in itself say what it is a digitization or a binary of.

Digital technology thus introduces a radical split between the character sign and its meaning, which can only be established by means of an arbitrary reading agreement: in practice, these agreements are formats such as JPEG, Unicode and ASCII, which associate content with a sequence of binary units.

This is why we speak of code: not in the sense of programming code, but manipulation code. In other words, we code content in the form of binary units (or units borrowed from any formal alphabet in the sense of a Turing machine) that make it possible to manipulate through calculation. This coding operation translates the content into non-semantic units, with only formal characteristics associated with their typing and distinguishability. There is therefore no direct link between these units and any meaning we might attribute to them. In such a coding paradigm, understanding a code is tantamount to knowing how to execute it, and thinking is tantamount to calculating. The problem of interpretation therefore needs to be tackled afresh.

To know	To understand	To think
Resemblance	Reveal	Contemplate
Causality	Reproduce	Explain
Codification	Execute	Calculate

Table 4.1. *Knowledge paradigms and epistemic operations*

4.4. Interpreting coded content

How is coded and calculated content interpreted? In general, a calculation is an intermediate step in deducing a result from the information coded for the calculation. It would therefore be possible to consider interpretation from the origin of the codes, on the one hand, and according to the context in which the results will be used, on the other hand. In other words, between origin and destination, the calculation is simply a stage where meaning is unnecessary for its execution, but not absent. After all, we all know what we are doing when we calculate. This vision, which seems natural, is not, however, tenable. The reason for this is that calculation breaks the link to the origin and does not determine any particular destination.

One way of looking at calculated codes from their origin would be to give them traceability: they would be the formalized residue of entities with material context and anchorage. A trace is what remains of a past event, a bygone past, whose non-contemporary nature is affirmed in the present environment. The code, however, breaks all ties with the past, making it possible to purely manipulate the coded content. While the trace resists the arbitrary interpretations we would like to see in it, given that it retains enough of its origin to constrain what we can make it say, the code, on the contrary, offers zero resistance to any interpretation (it has zero semantic inertia), provided that it is formally isomorphic – if we allow ourselves this pleonasm, since isomorphism is nothing other than the coincidence of structural forms. This is why coded content is not a trace, but a "given", what we establish or decree, what we give ourselves in view of the calculation to be performed. To calculate is to assume a break with the past, to establish a new situation, a formal configuration made up of meaningless symbols that enable a calculation to be carried out.

There is therefore a gap between the origin of the information under consideration and the calculation we derive from it. Assuming and managing this discrepancy requires the establishment of a rigorous interpretative agreement, in which we see how the calculation is no longer the origin, but an approximation, an idealization and a new conformation. Science is accustomed to such a practice: it is simply a matter of producing an operable, calculable model, and not confusing the model with the reality described.

Similarly, if the link to the origin is overstretched by the calculation, the relationship to the destination remains obscure. This is because a calculation has no goal; it merely consists of producing a result that is in itself neutral and devoid of any interpretation. Calculation creates an exception, a split between origin and destination, a timeless bubble with no past or future. We do not know where it comes from, nor where it is going.

Faced with such an interpretive gap, we need to develop a bypass strategy to understand what this calculation means. In most cases, science has developed a rigorous interpretative discipline in which the calculation is derived from models, which themselves establish the link with a represented reality. The problem of calculation is extended to that of modeling. However, this situation remains uncommon, if not rare. Many calculations are carried out without any explicit model. For example, massive data processing (Bachimont 2016) simply collects data, without examining it in terms of its relationship to reality. In practice, what is collected is what can be collected, and the aim is to establish links or correlations in order to derive knowledge about what these data are about.

However, there are two problems. First, the calculation performed has no direct or obvious connection with the nature of the data and their interpretation. The calculation exploits the formal properties of the data, whether they represent this or that. Second, the complexity of the processing, which has become quite opaque due to the blind nature of the data, makes it difficult to understand how the results produced are meaningful.

If we cannot interpret a calculation from the way it is produced or the data from which it is derived, then there is but one way to interpret calculations and their results, and that is storytelling, in other words, using a constructed narrative to give meaning to the data from their inception (once upon a time) to the revelation of the results (the outcome). This narrative is immediately intelligible, because it uses a form of communication that is unique to human beings, enabling us to rationalize what happens to us. If we can understand it, we need to know how much value to give it. If the latter cannot come from the treatment itself, which may be unintelligible in the absence of rigorous modeling, it must be granted on the basis of what we recognize in the results from knowledge we already had. Reiterating Meno's paradox (Plato 1999), we can only learn what we already knew, because if the result is indeed new, we cannot recognize or detect it as such.

This narrative is therefore interpretative, for lack of a better solution, arbitrarily applied to the calculated results. To overcome this, it is not necessary to give up on storytelling, as it remains an effective and efficient means of putting data back into the context of their origin and destination. However, to overcome the arbitrary nature that such narration might entail, we need to frame the role of data in a rigorous, context-specific modelization that articulates the formality of computational rules with the relationships considered in the real world.

There are then multiple tasks that can be distributed. If the creators of the calculation are in charge of proposing adequate modeling, then users can trust and wonder about the narrative justifying the origin of the data, giving relevance to the use of the calculated results. In the same way that we are satisfied using certain artifacts without worrying about their inner workings, calculated models are subject to the same fate: once validated, they are used without questioning their intelligibility or authenticity. We can conclude from this that the challenge now is to link data modeling with its computational implementation, so as to authorize a meaningful narrative that plunges the data back into a reality from which they had to be extracted and cut.

4.5. Conclusion

Digital technology has become the main tool used by our intellectual techniques. Based on calculation, it establishes a special relationship with reality insofar as it requires us to code it in order to make it manipulable. A code, however, is a set of meaningless formal symbols whose relationship to the represented reality is arbitrary and unmotivated. The same code, the same calculation, can refer to multiple incommensurable realities. Therefore, if representations of reality are the result of its coding, these representations no longer contain any intrinsic link of their own, enabling them to be articulated or connected to the reality they represent. From this point of view, digital technology represents a split from our traditional conceptions of knowledge: ancient knowledge, where knowledge is knowledge by virtue of its resemblance to the known it considers, and classical knowledge, where knowledge is the result of a causal process.

This arbitrary coding establishes and decrees data in a formal way that makes them calculable, independently of the constraints inherent at its origin: data are what we give ourselves, what we set and define. The only meaning we can associate with the latter is their manipulability and operationality: the meaning of data is reduced to the calculation rules we apply to them. Under these conditions, it is difficult to know how to interpret the result, except in a variable and arbitrary way.

This arbitrariness is not a defect. It is the prerequisite for universal calculations (as long as we know how to code, and abstract ourselves from the initial context, the algorithm is applicable) and can be automated (calculations without semantics can be carried out by machines). In order to overcome the difficulty caused by the arbitrary nature of the calculation, scientific culture has developed the notion of models, which rigorously establish the link between a context, the data that characterize it, the calculated results and the conclusions that can be applied to it. In many cases, however, we find ourselves using data and calculations without any associated modeling: only storytelling can therefore give meaning to the results and make sense out of them. Such storytelling, which cannot be controlled from a model, is only made possible by our prior knowledge or our desire to rediscover what we already think, at best seeing a confirmation in these calculations, and at worst a fantasy paired with the objectivity of calculation.

To escape this Meno's paradox, it seems essential to be able to link data modeling to the narration of its use. If only a narrative can tell the story of the origin and meaning of the "here and now" of the use of results, then semantic control of the narrative must be based on modeling, introducing semantic inertia into the data, making it impossible to make them say anything, even if the arbitrary nature of the code cannot prevent this. Correct calculation for the telling of relevant stories forms the new union of numbers and letters.

4.6. References

Bachimont, B. (2016). Traces, calcul et interprétation : de la mesure à la donnée. *Azimuth: Philosophical Coordinates in Modern and Contemporary Ages*, IV(7), 13–36.

Copeland, B.J., Posy, C.J., Shagrir, O. (eds) (2015). *Computability: Turing, Gödel, Church and Beyond*. MIT Press, Cambridge.

Daston, L. and Galison, P. (2012). *Objectivité*. Les presses du réel, Paris.

Gingras, Y., Keating, P., Limoges, C. (1998). *Du scribe au savant: les porteurs du savoir de l'Antiquité à la révolution industrielle*. Les Éditions du Boréal, Montreal.

Guenancia, P. (2000). *Lire Descartes*. Gallimard, Paris.

Hamou, P. (2002). *Voir et connaître à l'âge classique*. PUF, Paris.

Lévy, P. (1990). *Les technologies de l'intelligence. L'avenir de la pensée à l'ère informatique*. La Découverte, Paris.

Pasnau, R. (1997). *Theories of Cognition in the Later Middle Ages*. Cambridge University Press, Cambridge.

Plato (1999). *Ménon*. Canto-Sperber, M. (trans.). Garnier-Flammarion, Paris.

Pratt, V. (1995). *Machines à penser: une histoire de l'intelligence artificielle*. PUF, Paris.

Wagner, P. (1998). *La machine en logique*. PUF, Paris.

5

"Motion" Machines and "Token" Machines: Milestones in the History of the Alphabet

5.1. Introduction

In the following pages, we would like to reflect on the relationship between technology as a capacity to act and produce, on the one hand, and language as a capacity to symbolize, on the other hand. Within the French epistemological tradition, Leroi-Gourhan revisited this matter when he published *Le geste et la parole* (1964) ("Gesture and Speech"), the first volume of which, "Technology and Language", sums up the problem in its very title. However, for reasons we will explain later, this is not the source we start from, even though we recognize that the book in question marks an important stage in the questioning of the relationship between technology and language. Instead, we would like to start by drawing on the work of a later anthropologist of technology, François Sigaut, and in particular two comments he made in what turned out to be his last book: *Comment Homo devint faber* (2012) ("How Homo became Faber"). Based on these two observations, we then take a look at the distinction between "motion" and "token" machines, seen as milestones in the history of writing.

Chapter written by Jean LASSÈGUE.

5.2. Two comments on technology from François Sigaut

Which two points are highlighted by Sigaut? He presents both of them in a negative light, indicating that it is the opposite of the generally accepted view.

First, *tools are not language*. The potter's wheel, the hammer or the arrow are not, strictly speaking, intended to speak: they are *mute*. Technology is therefore not intended to be directly expressive (even if an expressive aura surrounds technical objects, particularly when they are destined for war), insofar as it does not primarily target the realm of meaning, but that of action and production. This is why the tool is usually limited to ensuring the sequence of cause and effect, and thus succeeds in making people conceive of a reality that is external to them, a reality that resists them and does not depend on their good will.

Second, *technical activity does not arise from a tendency that is already present in the individual*, as if it were merely multiplying or reinforcing an innate ability. For example, pliers are not an artificial extension of the hand: on the contrary, technical activity results from a *collective attitude that materializes in differentiated, synchronized gestures, forming styles*. From this point of view, technical activity does have something to do with meaning, since it presupposes a collective construction, even if the tool itself is not aimed at expression, but at action and production, as previously mentioned. These two comments show that the relationship between technology and language is more complex than it might appear at first, since they go in opposite directions: the former dissociates technology and linguistic meaning, while the latter brings them closer together, since technology is the result of a collective construction of meaning. This has two consequences.

First and foremost, *the technical attitude consists of imprinting certain ranges of gestures on bodies*, with a view to achieving collective goals oriented toward action or production. Sigaut takes up the notion of "body technology" developed by Mauss (1937) ("Les techniques du corps") in a famous article: "body technology" consists of the collective organization of bodily gestures to perform certain activities (weaving, sowing, hunting, etc.) and passing on these styles through generations. The tool is therefore not the first, but rather the ultimate result of collective gestures that precede it, even if the tool, once created, participates in the shaping of the gestures in question, which in turn never cease to evolve.

From this, we can deduce that the technical attitude *does not presuppose the existence of a tool.* A tool may appear "at the end of the line", as it were, as a precipitate of collective gestures, but first requires the collective organization of these gestures. This tool "at the end of the line" consists of *converting* the gestures informing bodies in *movement* in space by means of a tool that belongs to the world of "mute" causality, the first point made by Sigaut. There is nothing obvious about the transition from body gesture to movement in space, because they do not belong to the same phenomenology: the body itself *inhabits an environment with other bodies*, while movement *occupies a space.* The two orders are not immediately connected, even if the tool may eventually achieve this connection. Once established, the tool is both expressive in its semiotic construction and physical in its causal role.

The two points emphasized by Sigaut make it possible to set aside certain problems in the philosophy of technology, while simultaneously renewing the relationship between technology and language.

5.3. Renewal of the technology–language relationship based on François Sigaut

There seems to be two important points in this renewal. The first is negative: it allows us to discard an age-old tendency in the philosophy of technology to interpret the technical attitude as an extension of the individual body, neglecting its social aspect. This point of view can be found to varying degrees in the 19th century with Kapp (1877), who speaks of technology as a projection of organs onto the outside world, and in the 20th century with Leroi-Gourhan, when he speaks of the externalization of memory (Leroi-Gourhan 1965), as well as in the 21st century with Stiegler, when, following Alfred Lotka, he makes "exosomatization" one of the keys to humanity's technical relationship with the world (Alombert and Krzykawski 2021). Objections to these comparisons may well be raised by pointing out that the contexts in which these authors think differ from one another, but we will not delve into the details of each of their theories here: we only want to highlight what seemingly appears to be a profound historical trend. Sigaut radically stands out from this historical trend: by relating the technical attitude to a collective attitude, he does not consider technical activity from the perspective of the natural individual, but makes it a primarily social attitude that has effects on individuals via the learning of the stylization of gestures and, potentially, the construction of tools.

To us, this represents a major step forward in thinking about technology, insofar as the relationship between technology as a capacity for action and production, on the one hand, and language as a capacity for meaning, on the other, becomes *connected*: technical activity depends on a collective construction of meaning, prior to the realization of the technical activity itself, which is confined to the range of causality. There is nothing "natural" about it, as the tendency to extend it beyond the body itself still suggests.

Sigaut's point of view also seems to allow us to explore a question that he did not directly address, and which we can formulate in the following way, based on his framework of thought: if we acknowledge that language is a stylization of learned gestures arising from a collective attitude of "body technology", what does it mean, in this case, to *equip language with tools*? If we assume that language is the result of an analysis in terms of "body technology", writing would be one of the first tools to crystallize this body technology: this is what we are going to talk about here.

The question of the technical equipment of language thus appears as a counterpoint to Sigaut's question, which consisted of trying to clarify the semantic conditions of possibility of technical activity. The next step is to study the technical conditions of speech in writing.

5.4. Writing as a tool

We mentioned above that we could think of a tool as an object which converts the gestures that inform bodies into movement in space. From this perspective, writing would convert the invisible vocal gestures of speech into visible traces of movement in space. Writing would have two sides: it would be both a *learned gesture* when related to its bodily source and a *trace* when considered solely as movement in space. In this way, writing would play a part in both the world of meaning, insofar as it aims to capture speech as a gesture, and in the mute world of causality, insofar as it leaves a trace in the physical world. This link between gesture and trace, made possible by the tool, is delicate: the notion of the machine allows us to define its limits.

5.4.1. *"Motion" machine hardware*

When we think of machines, "motion" machines are the first thing that come to mind: these are hardware devices that Aristotle described as "artificial" (crane, pulley, wheel, potter's wheel, etc.) in the *Mechanics* (see De Gandt (1986)) because they struggle against the natural force of gravity (seeking to reach its "place") by imparting an external force to the machines counteracting this gravity. The machines in question are not autonomous, since they require an external force to artificially generate motion in the mechanical device. Increasing the autonomy of "motion" machines means attempting to reduce their artificial nature, so that they are endowed with as much natural motion as possible, such as the force of gravity. The external force used to move machines comes from any means to generate motion: the natural force of the elements (water, wind), domesticated animals or enslaved humans. The question of machine autonomy, that is to say, the relationship between natural and artificial movement, assumes a complex social organization geared toward the maximum naturalization of movement, its complete autonomy.

5.4.2. *The semiotic mechanism of "token" machines*

The question of the autonomy of "motion" machines bounces back when we consider another type of machine, which we call a "token" machine. This is another type of machine that not only aims for autonomy, *but ultimately achieves it in full*, and these machines *are directly related to writing in its alphabetic form*. We would now like to try and clarify two points concerning the alphabetic form of writing (we are using the term *alphabet* here in a broad sense, covering both the "abjad" and the "alphabet" itself, as we will see below): on the one hand, the "token" machine in question consists of tools based on the *combination of a finite number of elements representing speech elements*; on the other hand, there is a long-term historical trend that seeks to achieve the *full autonomy* of these "token" machines by dissociating sound from meaning in the act of reading and writing signs. This is an almost exclusively Western history, if we take the "West" in its broadest sense: from ancient Mesopotamia, Elam, Egypt and Greece to Europe and its colonial extensions, in other words, the alphabetical cultural zone. However, there is one notable exception: Korea predates the invention of the metal movable-type printing press by at least 50 years, which is generally attributed to Gutenberg around 1450. The Korean case is particularly interesting: in addition to chronologically being the first; it brought about the appearance of an alphabet, Hangeul (Korea Foundation 2010; National Hangeul Museum 2015; Réunion des musées nationaux 2018), conceived as a "token" machine, and the mechanization of this alphabet because of a "motion" machine, the printing press, all in the same historical moment. Chronologically, we can thus see the superposition of the "token" and "motion" machine in Korea: this is a unique example in history, to our knowledge, if we consider that 3,300 years separate the appearance of the alphabet from the appearance of printing in the West, taking the appearance of the alphabet in Sinai (1850 BCE) and Gutemberg's (1450) invention as a reference point.

Let us begin by characterizing the "token machine" that is alphabetic writing. The alphabet is a repertoire of a finite number of combinable elements designed to capture certain features of speech, which vary according to the graphic systems employed. The alphabet assumes a distinction between sound and meaning, and it is the modalities of this distinction that account for the variability of graphic systems.

It can be hypothesized that the "body technology" behind the alphabet consisted of fully separating the first and second articulations of language, as described by Martinet (1967), in other words, separating the morpheme level (carrying meaning) from the phoneme level (devoid of meaning). The alphabet isolates the two levels of articulation, and the history of the alphabet can be interpreted as a tendency to make the phonemic level of articulation increasingly autonomous from the morphemic one, that is to say, independent of meaning. This autonomy of the phonemic level of articulation means that the alphabet does not behave *like* a machine; it *is* a machine: *phonemic elements need nothing external to themselves to enable vocalization*, and over the course of history, this phonemic performance depends less and less on an external force, that of the construction of meaning that defines the level of the first morphemic level of articulation. In short, characters have come to the point (we describe the major stages in this history below) where they no longer need any interpretation to be read, but can be scanned without any external recourse – they have become mechanical.

In the West, the complete autonomy of the alphabet as a "token" machine has been the result of a long history and has not been simultaneously achieved in reading and writing. These are the two points that we would like to examine now.

5.4.2.1. *Progressive reading automation*

The separation of morphemic and phonemic levels of articulation, which was made possible by the alphabet, was progressive. From this progressive separation, we would like to mention three fundamental stages: the *abjad*, the *alphabet* and *code*. The word "abjad" designates the first letters of the Arabic alphabet and corresponds to the order of letters in the classical tradition, just as the word "alphabet" designates those of the Greek alphabet, whose order has been maintained to the present day. We would also like to point out that these three historical stages do not tend toward the last one, which would be like the final synthesis, and we should instead think of these stages as the formation of geological layers that are never really eliminated, that continue to act in their own way depending on the context. The generally tri-literate roots of Arabic represented in the abjad continue to impose themselves on the alphabet, as shown by the example of tri-literate acronyms, ranging from political parties to brands: PKK (Kurdistan Workers' Party), MBK (the former Motobécane bicycle brand), BRKLYN (clothing brand, a fine example of the use of a long vowel (Y), the only vowel category used in the abjad). Similarly, the Latin alphabet continues to be used in code, generally to write down English. These three stages correspond to three well-defined functions: the formation of a finite repertoire of signs enabling the sound/meaning distinction (*abjad*); the formation of a finite repertoire of signs enabling the automation of reading (*alphabet*); the formation of a finite repertoire of signs enabling the automation of writing (*code*). We describe them briefly below.

5.4.2.1.1. Abjad

This is the first alphabetic (in the broadest sense) writing system, the first traces of which have been found in Sinai and date back to 1850 BCE, if we refer to Goldwasser (2011). It consists of notating the consonants of words in order to reconstitute the morphemes of the language, in other words, the first level of articulation. Compared with the hieroglyphic, non-alphabetic writing that preceded it, this enabled a drastic reduction in the number of characters to 22.

Abjad is already alphabetic, but inherits earlier syllabaries with the aim of reconstituting the language's morphemes. By becoming ambiguous due to their small number, the written letters become unpronounceable, and it is the speaker who, through their active intervention in the discourse, removes the ambiguity of their composition by adding the appropriate vowels, thereby reconstituting the morphemes. Prior knowledge of the language and a relationship with the spoken language remain essential: they are external to the abjad, to the repertoire of consonants. Writing thus presupposes the reader's knowledge of the spoken language, since one *must already know the meaning of the word* in order to be able to place vowels in the right place and reconstitute morphemes: in short, one must know the spoken language in order to be able to read it.

From this point of view, abjad is very much like the machines that need an external force to be moved, as described by Aristotle in *Mechanics*: if you do not know the meaning that acts as the external force that sets the letters in motion, if you do not know how to put the vowels that recompose morphemes in the right place, you will not be able to read or write. The next step is to reduce the role played by this external force, and in the case of reading, to almost remove it altogether.

5.4.2.1.2. Greek alphabet

This is the alphabet referred to when distinguished from the abjad. It is Greek in origin, as its name suggests, and dates back to 800 BCE. It has a long lineage in Latin, Etruscan, Cyrillic and other alphabets. The Greeks did not invent the principle of alphabetic writing, that is to say, the notion of a repertoire of a finite number of combinable elements, but received it from the Phoenicians, with whom they had commercial contacts. However, the Greeks revolutionized the way this graphic system was conceived: it no longer recorded morpheme roots that had to be recomposed by adding vowels, but rather the sounds of the phonatory system, without taking morphemes into account. The Greek alphabet is therefore at a phonological deciphering level that is absent from earlier graphic systems, all of which retain a link with the semantic level. This is why, unlike the abjad, the principle of reading with the Greek alphabet does not require prior knowledge of the language: the letters note the phonemes and can therefore be *processed automatically*. We experience this every day when we are tired and manage to read a page without understanding it at all: we simply scan it automatically.

The Greek alphabet automates reading: from this point of view, the Greek alphabet is a *machine*, a *reading* machine. It is autonomous in reading, because it contains all the elements needed for deciphering, and does not require the reader's contribution, since there is no longer any need to reconstitute morphemes, only to scan phonemes.

5.4.2.2. *Writing automation*

The appearance of code in the 20th century, in the form of programming languages, has been innovative in that it aims to partially automate writing, rather than just reading. Characters are concatenated according to logical rules that require no interpretation and can be executed automatically. Several steps were necessary to achieve this result, two in particular. The first arose from an issue with space with the emergence of non-Euclidean geometries, which forced mathematicians trained in the formal tradition to strictly adhere to writing and the character sequences it makes possible. In his article "Über das Unendliche" ("On Infinity") (Hilbert 1926), Hilbert points out that it only presupposes the "[...] existence of certain extralogical concrete objects which, as immediate sensations, precede all thought. [...] With particular regard to mathematics, the object of our study will therefore be the concrete signs themselves, whose form we know how to distinguish and recognize from the point of view we have adopted". In other words, this is a question of sticking to purely logical necessity *outside space* that ensures the coherence of a formal system, its lack of contradiction. Hilbert's wish concerning writing was to completely separate the phonological anchor from its morphemic base, only retaining the formal combination of characters from the alphabetic principle. It would then be possible to stick to the formal sequence of characters from which to mechanically generate all valid propositions, whether they be arithmetic or geometric. To make such a system of tokens work, however, Hilbert still needed to postulate an external activity of the subject, as he points out in his comment in his article "Die logischen Grundlagen der mathematik" ("The Foundations of Mathematics") (Hilbert 1923): "[...] our thinking is finite, when we think, a finite process takes place".

It is this external finite principle, which Hilbert assimilates to thinking in general, that Turing succeeds in showing could itself be written in the form of a program executable on a machine (an all-abstract machine, the so-called "Turing machine"). The formal sequencing of meaningless tokens can thus be delegated to a program that can be run on a read and write machine: the computer. Given a list of instructions (in the form of tokens in a certain order) and data (in the form of tokens in a certain order), it is possible to process the tokens, in other words, to rewrite them: such processing therefore amounts to executing rewriting rules. From this perspective, the program writes itself, because it no longer needs to involve an external, finite principle of thought.

This radicalization of automation, which now not only affects reading but writing as well, emphasizes the feeling of intense dispossession that has beset us since the advent of digital activities: *we are read* by machines that exploit our digital traces and these machines *write down* our "choices" in advance. The evolution of the alphabet has thus taken automation to the extreme, gradually splitting the concept of the sign in two and distinctly separating sound from meaning. Ultimately, the computer appears to be the tool that is materializing this split, which arises from gestures that were collectively adopted in the distant past. The effect of the computer on all graphic systems is remarkable: even Chinese ideograms must be based, at a certain level of computer compilation, on a binary decomposition of their characters and on the notion of a program, which borrows its logic from the alphabet. From this point of view, the extension of the alphabet through the computer affects all graphic systems, even if users are more often than not unaware of it.

5.5. Conclusion

If we now take another look at the question of the relationship between technology and language, we can make three observations. First, the growing distance between the two sides of the sign – its morphemic reality, on the one hand, and its phonematic reality, on the other hand – appears to be just one of several anthropological options that have taken shape over the long term in the West. Second, while the extension of alphabetical order is well established and affects all graphic systems when they are computerized, this does not mean that the collective gestures from which these graphic systems originate have been definitively dismissed in speech: Chinese speakers know this, even though they continue to speak Chinese despite the alphabetical computerization of Chinese writing. Ultimately, the autonomy achieved by "token" machines over the course of history, from the alphabet to the appearance of the computer, leads us to believe in the naturalness of the latter: hence the misleading idea of an "artificial" intelligence that would have eventually or will eventually (the temporal hesitation is symptomatic) become "natural", whereas the computer, as a tool, only manifests autonomy in the recording of tokens, but not in the recording of movement. To believe that "motion" machines could behave in every way like "token" machines would be as naive as believing that nature is structured like a movable-type printing press. The world's readability is far from being equal to "token" machines, because the autonomy of natural systems remains an open problem: writing allows us to think about it, but not yet imitate it.

5.6. References

Alombert, A. and Krzykawski, M. (2021). Vocabulaire de l'Internation, Introduction aux concepts de Bernard Stiegler et du collectif Internation. *Appareil*, 56. doi: 10.4000/appareil.3752.

De Gandt, F. (1986). Les mécaniques attribuées à Aristote et le renouveau de la science des machines au XVIe siècle. *Les études philosophiques*, 3, 391–405 [Online]. Available at: https://www.jstor.org/stable/41582298.

Goldwasser, O. (2011). The advantage of cultural periphery: The invention of the alphabet in Sinai (circa 1840 B.C.E). In *Culture Contacts and the Making of Culture, Papers in Homage to Itamar Even-Zohar*, Sela-Sheffy, R., Toury, G. (eds). Tel Aviv University, Tel Aviv.

Hilbert, D. (1923). Die logischen Grundlagen der mathematik. *Math Annal.*, 88(160), 151–165.

Hilbert, D. (1926). Über das Unendliche. *Math Annal.*, 95, 161–190.

Kapp, E. (1877). *Principes d'une philosophie de la technique*. Vrin, Paris.

Korea Foundation (2010). Hangeul, Korea's unique alphabet. Seoul.

Leroi-Gourhan, A. (1965a). *Le geste et la parole, tome I: technique et langage*. Albin Michel, Paris.

Leroi-Gourhan, A. (1965b). *Le geste et la parole, tome II: la mémoire et les rythmes*. Albin Michel, Paris.

Martinet, A. (1967). *Éléments de linguistique générale*. Armand Colin, Paris.

Mauss, M. (1937 [1968]). Les techniques du corps. In *Sociologie et anthropologie*. PUF, Paris.

National Hangeul Museum (2015). History of Hangeul. Seoul.

Réunion des musées nationaux (2018). Caractères d'Asie. Trésors de la bibliothèque. Paris.

Sigaut, F. (2012). *Comment Homo devint Faber*. CNRS, Paris.

6

"Digital Technology", Revealing Intersections between Epistemology, Political Philosophy and Philosophy of Technology

6.1. Introduction

Taking a closer look at technology calls for humility: we come across inventions and human expertise in various centuries and places. Dictionaries explain that we must reconcile artistic gesture and repetition, and that, since Galileo, it has been difficult to think of technology without science[1].

Body technology (walking, swimming) (Mauss 1983)[2] reminds us that culture also refers to society, and that our daily lives are imbued with rhythms, social habits and *learned behavior*. This last expression is the definition of culture by anthropologist Ruth Benedict. This brings us close to a paradox: culture – defined as erudition or as the sign of a wider group – is often presented as being antinomic with technology (Jeanneney 2005). Malinowski has also been criticized for his functionalism (Malinowski 1968). Does sociology resolve this antinomy? The latter sometimes examines how technology is used to maintain or invent forms of

Chapter written by Éric GUICHARD.

1 See: https://www.cnrtl.fr/definition/technique.

2 Mauss calls technology "*an effective traditional act.* [...] There is no technology and no transmission if there is no tradition" (p. 371).

enslavement (Casilli 2019). However, with a few exceptions (Durkheim, Bourdieu and Chamboredon), its fields rarely cross paths with technology.

Technology also includes politics. Not just as a practical solution to its "service" ("social distancing"), but also as its vision: the focus on the determinism of innovation ("new technology will transform society") has been nurtured by most governments for a century, Edgerton tells us, for whom the reference to new technology is part of a narrative that is essential to our societies (Edgerton 1998): a myth that nurtures our "cultures".

Ultimately, epistemologists rarely make the link between technology and knowledge, even if they are invited to take an interest in technology (and institutions) by science historians (Pestre 2013). They are pushed into discretion by sociologists of science[3], who produce vast summaries that are incompatible with the role of human singularity in the construction of science[4]. They are also excluded from the debate on culture by those who consider themselves its owners (Dhombres 2022; Guichard 2022).

The question of technology thus intersects the fields of science and art on the one hand, and the social sciences on the other hand; for the latter group, it nevertheless appears secondary, and is even viewed as a *pariah*. Its relationship to knowledge and epistemology depends on the balance of power between the disciplines that address questions of science. Thus, bringing together epistemology, philosophy of technology and political philosophy (or even sociology) seems like an unrealistic task. However, using "digital technology" as a starting point, we attempt to show that this is not the case. This leads us to explore five theses:

– our thinking is essentially technical and always has been;

– writing is technical, and mastering it is essential to understanding the world;

– the Internet and digital technology are transforming writing;

– writing is being stolen from us;

– the focus of political philosophy on countries, institutions and citizens neglects the political role played by large digital companies.

---

3 Including Latour, criticized by Bourdieu (2001).

4 A Fourier or a von Neumann machine is difficult to fit into a sociological trajectory, even if they may have interfered with society and politics.

6.2. Our thought is essentially technical

We have, since early times, used machines to count and used angles[5] to build. Ptolemy devised his system of cartographic projection based on the angular measurement of a point on the Earth in relation to a few stars (Desbois 2015). Galileo's telescope, like the Webb telescope, reminds us that we have been surveying the cosmos for thousands of years in order to learn more about our environment, as well as to develop our representations of the world and ourselves.

Physicists make great use of instruments and measurements. For the humanities, the Greek characters sculpted and cast by Claude Garamont[6] have become the Galileo telescope for specialists in Antiquity.

Mathematicians have tried to drive the function of tools and machines in thinking and demonstration to its limits. While Gödel (1931) demonstrated that it could not be comprehensive, Turing (1936) theorized in what conditions we could build a universal machine – the so-called Turing model. This led us to computers, which also equip our thoughts. Our intelligence has therefore been artificial since the dawn of humanity.

Our intention is not to enhance technology at the expense of the concept but to insist on their interweaving and on the multiple interactions between what we believe to be thought and materiality. The importance of expertise in the construction of science and proof is well-known[7], and a mental operation can be mechanical[8]. However, the fact that there is no sophisticated thought without technology is questioned by so-called "spiritualist" thinkers, who believe in an innate superiority of the mind over the objects we make (and who neglect the elementary operations of this mind). The followers of Husserl, Heidegger and Habermas have endeavored to convince us that technology is related to a *sin* since Galileo's discoveries (Guichard 2022); and as dictionaries tell us that it is the daughter of science, the latter is held responsible for many of our troubles.

5 With empirical knowledge of the Pythagorean theorem: three sticks of lengths 5, 12 and 13 for the Egyptians, or lengths 3, 4 and 5.

6 Called the "Grecs du roi" ("the king's greeks") and commissioned by François I from Robert Estienne. Their passage through Geneva, where Robert Estienne had taken refuge, generated considerable diplomatic tension.

7 See the dexterity of Coulomb and Joule (Pestre 2013).

8 Example: when we perform "mental calculations" – 17×5 is half of 17×10.

Granger has highlighted the importance of calculation in our intellectual operations. When we believe we are thinking in the common sense of the term, we are repeating and reproducing a large number of recipes of things we have learned, and thinking itself is their envelope: “For a thinking subject, the act of calculating involves non-mechanical aspects, possibly vicarious or adverse, but which certainly contribute to giving meaning to the process [...]. We have given the name ‘thought’ to these environments of calculation, which, in a certain way, overflow it” (Granger 2001, p. 76).

We hear that new (digital) technology is revolutionizing our mental tools. We have seen, with Edgerton, that the ideological aspect of such statements, as well as the idea of radical novelty due to digital technology, is untenable, as our thinking has always been instrumented.

6.3. Writing is a technology

To define writing, we usually evoke four components: a system of signs, one or more media, schools and academies, and ego (the singular subject). The interactions (relationships) between these constituents are at the origin of a set of underestimated abilities that are the foundation of the culture of writing.

The system of signs is fairly extensive (there are around 100 in France, not 26) and the media are variable: screen, hard disk, marble, etc. In all writing societies, there have been networks of scholars who debated the problems of interpretation raised by writing: writing reproduces language poorly (Olson 1998). The aim was to make the most of writing (for poetry, law, mathematics, commerce, etc.), to interpret and comment on what could be understood from it – or *should* be understood from it, in the religions of the Book – while refining it (without any idea of progress; there were numerous *scribal* regressions throughout history). Crosswords and sudoku testify to the solitary component of our relationship with writing.

Mastering the art of writing is a matter of learning, repetition, collective skills and quasi-mechanical rules. Writing is therefore a technology. It obviously helps us think: what would we do without the Internet, our pens and our books? This reality is obscured by the aforementioned spiritualism, and by focusing on its contribution to memory, which is nonetheless minor when compared to other functions of writing (see Chapter 5 by Jean Lassègue in this book).

The technicality of writing was poorly understood until Goody clarified it in a number of articles (from 1963) and in a seminal work (Goody 1979). He defined it as a *technology of the intellect.* Writing does not change society – to believe this would be technical determinism. It intrudes, transforms and nurtures new representations[9]. Simply put, writing is as much about the human as it is about society and becomes inseparable from both. Once writing is present, humans spend their time *writing* about their origins, their purposes, etc., and monotheistic religions are no exception. We can no longer refine our understanding of the world without "using" writing, whether it be in physics, medicine or philosophy.

Writing is therefore a tool that distorts the way we think about the world[10]. This fact presents a problem: if writing *acts* on our mental categories and representations, we have to admit that our tools are less objectifiable than we believe (Guichard 2017), and that the subject/object divide is less relevant than we claim it to be[11]. The technical/cultural opposition is also crumbling: how can we think, invigorate culture and bear witness to it without writing[12]?

Writing shows us the fragility of the oppositions between thought/technology, subject/object and technology/culture, which structure the way we think about the world.

Writing does not give us intrinsic power, it gives us abilities. However, in the face of a large proportion of people who cannot write, the group of literates[13] can be influential. This observation will be invaluable in

9 For an introduction to Goody's thoughts, see Goody (1986).

10 It is also a reflexive tool: its functions, effects and implications can be described by its use alone. This is a quality shared with language, which is also a "technology of the reflexive intellect".

11 Granger, quoted above, considers calculation to be virtual (as opposed to actual).

12 This point, which is valid for societies that write (and therefore today's societies, since all of them now write), has been spread to oral societies of the past through the work and testimonies of anthropologists and travelers – whose writings can, moreover, set the myths and traditions of the regions studied.

13 A "literate" person is defined as someone with strong writing skills in a particular field, which may be history, engineering, poetry, chemistry, etc. A literate person does not always

understanding the digital era. The "powers of writing", historically associated with literate worlds, have been extensively discussed (Baratin and Jacob 1996; Martin 1996; Goody 2000).

Despotic powers can threaten literates – see the ends of Giordano Bruno, Galileo and others. However, generally speaking, literates easily impose their moral, aesthetic and political values on the less literate[14], especially if they are few in number. Mastery of the written word thus gives the ability to participate in the "writing" of the world, at least through its representations. This fact refines the idea, which is common in the social sciences, that the world is *written* in favor of dominant groups – through law and history[15], for example.

Societies, however, are not static, and this writing of the world can be constructive, joyful and a source of social mixing. During the Renaissance (1450–1600, arbitrarily stopping at Plantin), goldsmiths, metalworkers, proofreaders and translators of dead and living languages and composers, rubbed shoulders with writers, theologians, dukes, bankers and mathematicians; Estienne's servants spoke Latin (Grafton 2007). Jenson, a goldsmith who had passed through Mainz, shined in Venice, and Manutius' proofreaders treated themselves to generous meals. Plantin welcomed the whole of literate Europe to his home. All these publishers engraved punches, molded typefaces, dried ink-dampened sheets and made books while correcting them. Physical effort was combined with intellectual effort, erudition with materiality, and the practices of one and all were emancipated from considerations of social class. The construction of culture is highly technical.

This will help us understand the interplay between heavy industry, precision craftsmanship, organization and the dissemination of thought. Writing refers to the intertwining of matter and the mind. It is sometimes in a cumbersome, noisy and messy context that knowledge makes great strides, where tolerance is forged that will serve as a melting pot for democratic ideas, agnosticism and atheism.

---

engage in reflective thinking: the sophisticated practice of writing does not always lead to knowledge being put into perspective.

14 And if there is a conflict of values, for example, political (e.g. between republicans and royalists, secularists and religious people), these are settled *within* the literate world.

15 It is also criticized for the lack of thoroughness in the notion of the social construction of reality (Hacking 2001).

Reimagined writing often leads to a renaissance[16], whether in Alexandria, Baghdad, Palermo (with Roger II and Al Idrissi), Venice, Paris, Geneva or San Francisco. However, there is nothing *written* in these intimate combinations of technology and culture.

6.4. Internet as writing

With computer science (and mathematics), we can produce any number and any sign with a two-term alphabet. Once we have demonstrated that it is possible to carry out certain operations (those that produce *calculable* numbers) with a machine that writes 0 or 1 on a ribbon from a finite number of elementary instructions and that we have equipped these machines with coherent grammar (Schützenberger and Chomsky's 1963 theorem), we can imagine that computer science has solved the question of the sign system. In actual fact, it has reopened it (Herrenschmidt 2007).

After all, we have to learn how to produce these signs and the rules associated with them, rather like in elementary school. Then we need to clarify what this writing potentially does to our representations of the world, to domesticate ourselves in it and to domesticate (regulate) its new actors.

The printing analogy still holds true: there are goldsmiths (who sculpt and melt signs, today's semiconductors), paper experts (supplemented by our hard disks and screens, our printers), grammarians (the inventors of programming languages) and authors. We can also see the multiplicity of interactions between theory and applications, conceptual and technical determination, innovations and reproductions at the origin of today's computers.

In practical terms, the Internet is also about writing: today, we are writing with our networked machines more than ever and we are constantly reading. This is all thanks to invisible blocks of text that guarantee communication: *protocols*.

This raises a question: who are the "digitally literate"? There are a variety of people in the world trained in all forms of computing and capable of pushing its limits. However, the expertise associated with this new culture of

16 For a balanced history of the Renaissance, see Goody (2020).

writing[17] is still not widely shared. A 10-year-old child can read a print or a 19th century manuscript, but a master's student is lost when faced with a text file that does not have the extension he or she expects and, like most adults, does not know how to "paste data" from a spreadsheet into a text editor.

Perhaps this is due to the fact that we do not take sufficient account of the a priori auxiliary culture of all writing, made up of minor methods and expertise that enable us to document, draw, produce graphics, write and organize our texts and so on. As a result, despite numerous publications on the subject, we do not have a good grasp of Internet issues, particularly in political terms (Bortzmeyer 2019): the reflexivity of writing can hardly precede its technical culture.

6.5. The robbing of writing and our free will

After the 1980s, the computer industry focused on building a monopoly over writing, complete with its privatization (formats, licenses, etc.). Even today, many people would be lost if we removed Word from their computer.

Admittedly, writing has always gone hand in hand with commerce: paper and fonts were sold, as were encyclopedias. While some claim that the massive use of Microsoft products has enabled the deployment of a culture of digital writing, we doubt it in view of the examples cited above about students, especially given our governments' taste for anti-reflexive "technical and commercial" solutions: by favoring machines made in China, by making it difficult to employ qualified people (secretaries included) and by signing agreements with companies to "bring schools into the digital era" (comments made by François Hollande in 2016).

The establishment of a culture of writing is also hindered by competition between "digital" companies, which encourages them to change their file formats every 5 years and hijack young Web standards[18].

17 Which Olson describes as the combination of technical expertise related to writing and the reflexivity fueled by these practices.

18 See: https://en.wikipedia.org/wiki/HTML for a modest description of the conflicts surrounding html5, back in 2011.

For the past 40 years, we have been living in an environment where writing is highly unstable and illiteracy is therefore widespread[19], amplifying our difficulties in reconnecting with rationality.

A turning point came in the 2010s: LibreOffice – a fork of the OpenOffice foundation – found an audience, Microsoft documented and clarified its Word format (docx), which Apple offered to make viewable, a priori, without a license, and Google offered free online writing tools. Visible costs are shifting toward infrastructure: we pay a high price to stay connected[20], so that we can write, save photos and communicate online: in the cloud.

There are also invisible costs: if writing informs our thinking, it is not difficult to understand that word processing and related tools[21] influence it. Our use of writing is also related to documentation, and we know that search engines direct it, by singling out their answers. The same applies to so-called social networks, between moral standards[22], message sizes and the syntax proposed for exchanges[23]. We write on private service platforms, according to the terms and conditions they offer. "Our thoughts no longer belong to us", writes Mathias (2009).

We are also distracted: by the advertising that invades our screens and those of our young children, by tools that claim to "advise" us (Siri from Apple). A wrong gesture on an iPhone screen and we are suddenly on a page with the stock market price, the addresses of nearby restaurants and the tennis champion of the day (tested on September 12, 2022). There is so much unwanted, oriented information. So-called social networks put us in a state of emotional overload, with filters we did not chose.

None of this is conducive to free will, serene thought or deep reflection, as it is often only displayed on tiny screens (an eighth of an A4 sheet). The

---

19 Which is not incompatible with the empirical expertise of certain political actors regarding so-called social networks.

20 For the purchase of a telephone and a network subscription, it is around EUR 70 for a single adult.

21 Numerous articles have shown how the use of PowerPoint stifles us intellectually.

22 Censored photos of bare-breasted women, promotion of hatred and racism. See: https://www.lemonde.fr/pixels/article/2022/09/29/massacre-des-rohingya-facebook-a-joue-un-role-central-dans-la-montee-du-climat-de-haine-en-birmanie_6143611_4408996.html.

23 Preference given to positive or simplistic ("like"), if not divisive, responses (Chavalarias 2022).

second level of writing (freely organizing our thoughts with it) is confiscated from us by those who have access to its storage.

Ultimately, the writing experts are selling us, not just to advertisers, but to political groups or dictators in order to influence us electorally and effectively: elections for Trump and for Brexit in 2016, in Bulgaria and perhaps in Sweden[24] in 2022. Our vote is not worth much in this new market: EUR 4,000 for a million people, says Chavalarias, who has documented this sinister business (Chavalarias 2022) of manipulating our free will in political affairs.

The power of writing can be seen in three grades: first, the 95% of us who are digitally illiterate find it extremely difficult to organize our thoughts; second, certain literates who have access to this storage of written words can blur our reasoning and guide our consumerist practices; third, they can manipulate our free will in favor of anti-democratic parties. The practitioners of such trades have, to our detriment, understood the link between the technicality of writing and inner thought.

We might think that all it would take is for citizens to become aware of the fact that their writing is being recorded and to prompt governments to reevaluate their vigilance and education, particularly in digital matters. Nothing is less certain. Indeed, these governments are now poorer than the digital industry.

In 2022, the French government budget oscillated between EUR 350 and EUR 500 billion (depending on whether we focus on revenues or expenditures). The gross domestic products[25] of Tunisia and France were, respectively, around EUR 40 and 2,500 billion. We know that this indicator is open to criticism, notably because it underestimates the role of public spending (Charolles 2022); but it does give an order of magnitude of a country's wealth.

Only nine countries have a GDP in excess of USD two trillion[26], and around 30 have a GDP in excess of USD 500 billion[27]. In comparison,

24 See: https://www.theguardian.com/world/2022/sep/04/sweden-high-alert-election-interference-amid-russia-tensions.

25 Among various definitions: "The sum of everything a country has consumed and invested".

26 At the time of writing, the euro and the dollar were equal.

Google's sales were expected to reach USD 260 billion in 2021, while its market capitalization was in excess of USD 1,500 billion. Google, which has no hospitals or schools to finance, posted a profit of USD 76 billion[28]. In the same year, Facebook posted sales of USD 118 billion and estimated profits[29] of between USD 20 and 40 billion.

Other writing firms handle similar sums of money. Taiwanese "funder" TSMC, which holds over 50% of the world's semi-conductor market, invests EUR 30 billion (or dollars) a year to "increase its production capacity and invest in research and development"[30].

Thus, the intellectual alienations and democratic threats mentioned above are organized or facilitated by companies with financial resources that exceed the GDP of half of all countries[31].

It is possible to consider that these concerns are exaggerated on the grounds of an alliance between these companies and wealthy states: critical economists and sociologists, who are often American, have shown that the United States has favored its enterprises for the past century in order to assert its hegemony (Masutti 2020). This point can be summed up by the notion of the *military-industrial complex* and the announcements made by Snowden, Assange and other whistle-blowers about the NSA's generalized data-filing of populations relating to the collusion of interests of a state and its big industry, or what is now called *surveillance capitalism*, at the expense of its citizens' freedom. This pattern, which works for states such as Russia and China, is considered relevant in the United States by libertarian advocates.

This analysis is also present in Europe, where the majority of governments have placed themselves at the service of industry and world

27 Random source (all are consistent): https://business-cool.com/decryptage/classements/pib-classement-pays-riches-monde-2022.

28 Google had more than EUR 100 billion in treasury cash in March 2022. See: https://www.webrankinfo.com/dossiers/google/chiffres-cles.

29 Five to 10 billion in profits per quarter by 2021.

30 One hundred billion euros (or dollars) between 2021 and 2024. See, among other sources, https://siecledigital.fr/2021/10/15/tsmc-annonce-la-construction-dune-usine-de-semi-conducteurs-au-japon.

31 The average GDP of the world's countries is USD 30 billion, which is a good estimate. See: https://fr.wikipedia.org/wiki/Liste_des_pays_par_PIB_nominal.

trade (Masutti gives harsh examples of ministerial advisors who have become Google communications advisors) and have favored the privatization of their education (see the aforementioned example of the contracts between Microsoft and the French Ministry of Education), at the risk of actively participating in the weakening of the states they are responsible for.

None of this encourages alternatives that promote experience, imagination and appropriation in terms of digital writing culture.

We can also consider that states no longer have the means to fight against the big digital companies, as the figures above demonstrate. The owners of these companies, with their colossal fortunes[32], have no hesitation in promoting political systems that do away with states, on the grounds that the algorithms of their companies will handle social issues better than states, including legal issues (Garapon and Lassègue 2018). Their arguments are the kind of primitive solutionism (Masutti 2020, pp. 341–345) that Musk is accustomed to, but their reach is real (they possess the communication systems that individuals, the media and governments use) – proof of their potential when it comes to *writing about world*.

We can finally take a more continuous view of the governing–governed relationship, noting that many French executives, who have been schooled in spiritualism (or ultraliberalism), have not helped in thinking through the emergence of the Internet (a fact that is obvious at university) and that our governments must adapt to the rising suzerainties of our countries (Supiot 2010): democratic states are not just the playthings of a caste that owns them. Europe is also the only political structure that offers protection to its citizens via the General Data Protection Regulation. Many of its members have developed political concepts based on a state with explicit duties toward its citizens in terms of law, social protection, emancipation and education.

32 Ten people have personal fortunes ranging from USD 90 billion to 219 billion. Among them, seven own or have founded companies closely linked to the digital world: Musk (Tesla, SpaceX, Twitter), Bezos (Amazon), Gates (Microsoft), Zuckerberg (Facebook/Meta), Page (Google), Brin (Google), Ballmer (Microsoft). See: https://www.forbes.fr/classements/classement-exclusif-milliardaires-2022-elon-musk-devient-lhomme-le-plus-riche-du-monde-devant-jeff-bezos-le-francais-bernard-arnault-en-troisieme-position.

To remedy this robbing of writing and its harmful effects, the ideal solution is to promote digital literacy to teach everyone (young, old, French, foreigners, etc.) the arts of writing, counting, drawing, documenting and reasoning with binary and reticulated writing. It sounds expensive but, on the one hand, the question of cost has never worked for writing, which has always seized enormous resources – even in ancient Egypt (Goody 1986, already quoted); on the other hand, "this is the price" for our ability to act and to develop refined analyses of the contemporary world.

Edgerton shows that a society made up of skilled technicians is better able to cope with technical innovation than one made up of illiterates surrounded by a few high-tech companies. Widespread mastery of digital writing could stimulate the kind of creativity seen in the Jenson, Manuce and Estienne networks, empowering uses that elevate us instead of dumbing us down. It is therefore better to invest in quality education than in ephemeral start-ups, based on the customer alienation model of the aforementioned companies. This is what Dennis Meadows mentions about climate education[33] when inviting us to break our habits, our "addictions".

Given the temporal regimes of digital writing, this will be possible over a decade or so, as we have seen.

6.6. Should political philosophy be renewed?

Can we take these new powers into account and imagine a political philosophy that can limit them? The latter facilitated the construction of nation states. As a result of the power relations between various social groups (which vary, but can be adequately characterized; see Marx), states that are committed to democracy have sometimes adopted segregative practices with regard to certain categories of their populations (immigrants and laborers in France, Jews in Germany, etc.). The result was a polarization of certain analyses, with the state on the one side and the citizen on the other side. Increasing our freedoms was seen as synonymous with combating abusive state prerogatives. Today, however, states, like their citizens, are dominated by more powerful structures.

33 Lecture given on the occasion of the awarding of an honorary doctorate to him at ENS Lyon on September 19, 2022. See: http://www.ens-lyon.fr/evenement/lecole/remise-de-doctorat-honoris-causa-dennis-meadows.

It would be beneficial to move away from this binary model that is still attested to by contemporary activists who claim to be devotees of left-wing ideas: emancipatory, with a collectivist and solidarity-based perspective. Of course, globalization – with its rare but sometimes positive consequences – has helped or forced us to broaden our political horizons.

However, if states no longer represent ultimate power and are becoming minor actors in an environment where others are already internationalized, then battles against their imagined or proven maleficence are meaningless if they do not consider the wealthy organizations that want to weaken or eliminate these states, which have the objective of further increasing their grip on our writing and our free will: the government, in its current form, is as fragile as our social protection systems, and all the more worth defending as it is the guarantor of our solidarity.

We therefore need to imagine a political philosophy that takes these three pillars into account: the citizen, the state and writing industry. The robbing of our common good, writing, by private structures with proven economic and intellectual power should make us realize that our freedom to decide our political future in the above-mentioned binary terms is obsolete.

Using figures has also offered us a reality check, as much appreciated by physicists (the importance of the order of magnitude) as by epistemologists: Damien (1995) showed, with Gabriel Naudé who collected thousands of books, how raw quantity forces us to rethink the organization of knowledge. This is proof, as Valérie Charolles reminds us in this book, that thinking about numbers and figures remains a worthwhile debate in philosophy.

In the first half of this chapter, we saw that the Internet is about writing and that writing is a technology: a trial addressed to the digital world will therefore have to extend to writing, which is an inconsistent approach. Furthermore, we know that judging technology is logically contradictory (Guichard 2017). Additionally, anathemas about modernity or industry are not very productive (Dagognet 1995).

On the other hand, it is always possible to judge contemporary individuals who are responsible for the criminal misuse of certain technology, all the more so when these misuses are aimed at undermining the political systems we have chosen – or at the callous exploitation of vulnerable people (Casilli 2019). Without ignoring the responsibility of mediocre opportunists who are intent on harnessing wealth or power and who also have unfortunate tendencies for hasty and collective judgment, it is important to strengthen our legal arsenal in terms of citizens' rights in the face of abuse by "digital" actors. In order to resolve these ethical and legal issues, it is possible to enlist the help of computer specialists. As digital literates, these people know how to spot contemporary appropriations, robbings and technical misuses, as well as their actors and perpetrators. They also get involved to promote freedom. Like the literate worlds of the past, their analyses and conclusions are diverse, and it is within the latter that the future is likely to be written. This encourages interdisciplinary dialogue.

6.7. Conclusion

Our starting point was an epistemological question: what is knowledge as a set of intellectual tools? We saw that writing is one of these tools, a technology that we have been shaping for 5,000 years and which, in turn, considerably informs what we call *thought*.

This brought us to the powers and capacities of writing. While the latter stimulate creativity, reflexivity and the search for well-ordered knowledge, the former can be considerable in terms of influence (aesthetic, moral, political, as well as economic). We have seen that the Internet (now known as the "digital world") is about writing, and it clarifies the functions of this digital writing. Since 2010, its appropriation has generated strong threats to countries that refer to themselves as democracies, as well as threats to their inhabitants, who have been reduced to the status of compliant consumers, even when it comes to voting. The intrusion of these new actors into the political field encourages us to imagine a political philosophy that integrates three pillars rather than two and to make the most of the reflexivity of writing by engaging in constructive debate with its contemporary literates.

6.8. References

Baratin, M. and Jacob, C. (eds) (1996). *Le pouvoir des bibliothèques*. Albin Michel, Paris.

Bortzmeyer, S. (2019). *Cyberstructure. L'internet, un espace politique*. C&F éditions, Caen.

Bourdieu, P. (2001). *Science de la science et réflexivité.* Éditions Raisons d'Agir, Paris.

Casilli, A.A. (2019). *En attendant les robots*. Le Seuil, Paris.

Charolles, V. (2022). *Se libérer de la domination des chiffres*. Fayard, Paris.

Chavalarias, D. (2022). *Toxic Data*. Flammarion, Paris.

Dagognet, F. (1995). *L'invention de notre monde : l'industrie, pourquoi et comment ?* Les Belles Lettres, Paris.

Damien, R. (1995). *Bibliothèque et État. Naissance d'une raison politique dans la France du XVIIe siècle*. PUF, Paris.

Desbois, H. (2015). *Les mesures du territoire. Aspects techniques, politiques et culturels des mutations de la carte topographique*. Presses de l'Enssib, Villeurbanne.

Dhombres, J. (2022). Quelques interrogations à propos de la construction cartographique. Effets culturels et contraintes mathématiques. In *Études digitales : cartographie et visualisation. Regards d'épistémologues et de concepteurs*, Guichard, É. (ed.). Garnier, Paris.

Edgerton, D. (1998). De l'innovation aux usages. Dix thèses éclectiques sur l'histoire des techniques. *Annales Histoire, Sciences Sociales*, 4/5, 815–837.

Foucault, M. (1969). *L'archéologie du savoir*. Gallimard, Paris.

Garapon, A. and Lassègue, J. (2018). *Justice digitale*. PUF, Paris.

Goody, J.R. (1979). *La raison graphique. La domestication de la pensée sauvage.* Éditions de Minuit, Paris.

Goody, J.R. (1986). *La logique de l'écriture : aux origines des sociétés humaines.* Armand Colin, Paris.

Goody, J.R. (2000). *The Power of the Written Tradition*. Smithsonian Institution Press, Washington/London.

Goody, J.R. (2020). *Renaissances. Au singulier ou au pluriel ?* Armand Colin, Paris.

Grafton, A. (2007). Vers une histoire sociale de la critique textuelle. In *Lieux de savoir – tome 1 : espaces et communautés*, Jacob, C. (ed.). Albin Michel, Paris.

Granger, G.G. (2001). *Sciences et réalité*. Odile Jacob, Paris.

Guichard, É. (2017). La philosophie des techniques revue à l'aune de l'internet et du numérique. In *Le numérique en débat. Des nombres, des machines et des hommes*, Chazal, G. (ed.). Éditions Universitaires de Dijon, Dijon.

Guichard, É. (2022). Pourquoi le rôle des sciences et techniques dans la construction de la culture est-il si peu évoqué ? In *Enquêtes en technosciences. Réflexions libanaises et françaises*, Salhab, M., Beaune, J.-C., Barbero, O. (eds). L'Harmattan, Paris, 55–72.

Hacking, I. (2001). *Entre science et réalité. La construction sociale de quoi ?* La Découverte, Paris.

Herrenschmidt, C. (2007). *Les trois écritures. Langue, nombre, code.* Gallimard, Paris.

Jacob, C. (ed.) (2007). *Lieux de savoir – tome 1 : espaces et communautés.* Albin Michel, Paris.

Jeanneney, J.-N. (2005). Quand Google défie l'Europe. *Le Monde*, 22 January.

Jeanneney, J.-N. (2010). *Quand Google défie l'Europe.* Mille et une nuits, Paris.

Malinowski, B. (1968). *Une théorie scientifique de la culture.* Maspero, Paris.

Martin, H.-J. (1996). *Histoire et pouvoirs de l'écrit.* Albin Michel, Paris.

Masutti, C. (2020). *Affaires privées. Aux sources du capitalisme de surveillance.* C&F éditions, Caen.

Mathias, P. (2009). *Qu'est-ce que l'internet ?* Vrin, Paris.

Mauss, M. (1983). *Sociologie et anthropologie.* Quadrige/PUF, Paris.

Olson, D.R. (1998). *L'univers de l'écrit.* Retz, Paris.

Pestre, D. (2013). *À contre-science.* Le Seuil, Paris.

Supiot, A. (2010). *L'esprit de Philadelphie. La justice sociale face au marché total.* Le Seuil, Paris.

PART 3

The Subject in the Era of Digital Metamorphosis

Introduction to Part 3

In this third movement, what has come to be known as the "digital revolution" is examined through the prism of "metamorphoses", which takes into account the future of technology by questioning its capacity to bring about brutal and structural changes in society. Indeed, it would not be difficult to show that every era is in fact a unique "technology era", and that subjects are as much its products as its actors. If the role and place of technology in society have been forgotten, it is also because subjects have incorporated and contributed to its changes without even being aware of it. However, the subject, which philosophy has often claimed to think of in terms of its purity or authenticity, has in reality never been anything other than a technical subject, changing as technology evolves. Our "digital era", characterized by the massive development of technology based on codes, algorithms and calculations, like Pascal's calculator was in its day, is not the result of a revolution, but rather of a metamorphosis that we, as technical subjects, have naturally embraced.

Its specificity lies in the multiplication of so-called "intelligent" technology, which produces an uncanny feeling in Freud's sense of the term: we are now less and less able to distinguish ourselves from this technology, which has had the paradoxical effect of making us realize that all living things are, in fact, a mixture of the living and the non-living. Awareness of "who" or "what" we are therefore immediately brings us face-to-face with our technology, which inhabits us as much as it surrounds us. Should we fear

Introduction written by Élise LAMY-RESTED.

the development of "artificial intelligence", which is still confined to executing programs and calculations, or should we, on the contrary, accompany it through its metamorphoses? How can we take care of ourselves as technical subjects, once we have understood that there is no such thing as "technical objects" completely external to us?

This section expresses and questions the concerns raised by the development of digital technologies, which we also call "new technologies" because they are algorithmically based (sometimes confused with discourse or *logos*), and the production of the discourses they generate. After a presentation of Bernard Stiegler's general organology, which proposes a restructuring of society as a whole, this part of the book is built around the dialectic between the desire to reactivate a truly human way of thinking, as argued by Anna Longo, and the risk-taking that involves accepting a form of machine autonomy, as well as the interplay between machine and human forms of intelligence, as defended by Tyler Reigeluth.

As a tribute to Bernard Stiegler and his desire to take better care of us, especially the younger generations and those in precarious situations, Vincent Puig's chapter considers the confusion between technical subjects and digital technology: by taking care of the latter, we are in fact taking care of ourselves. The story of his journey with Bernard Stiegler illustrates this ambition, in a world where technology currently remains at the service of economic and political dynamics that are blind to the ecological collapses they have caused.

Anna Longo's chapter, "Predictive Machines and Overcoming Metaphysics", questions the logician's dream of reducing all forms of thought to calculation, a dream that would find its place in "artificial intelligence". Playing on the opposition between Carnap and Heidegger, the logician defends the singularity of thought and its irreducibility to mere calculation. The author then, in her words, proposes "the creation of a technology that enables us to see the world as something other than a set of information patterns".

Lastly, Tyler Reigeluth's chapter, "Artificial Intelligence's New Clothes", suggests a form of human intelligence whose exercise would not be synonymous with a form of control and argues that such a perspective tends to maintain an asymmetrical power-knowledge pattern between an intelligent position that commands (that of the master) and an automatic behavior that executes (that of the slave). Without lapsing into ontological confusion, we need to reflect on how human and machine behaviors are articulated within social activities, and how one is the condition for the other's intelligence. If a machine is capable of learning, it will necessarily learn from its relationships. This implies rethinking the relational modalities of such learning, beyond any domination or automatic enslavement.

7

Taking Care of Digital Technologies with Bernard Stiegler

Thinking about technology in relation to time and spirit will remain Bernard Stiegler's great unfinished work[1] and will be an infinite source of questioning, in other words, of knowledge, for generations to come. This temporal and spiritual dimension is, for him, the condition of technology as a place for knowledge, the so-called "*milieu des savoirs*" ("milieu of knowledge"). Bernard Stiegler readily applied this definition of technology to the digital world (Stiegler 2014), a digital world whose organology (biological, technical and social organs) he first suggested studying in a general sense, so as to be able to consider its pharmacology. Indeed, any organ, whether biological, technical or social, can be a poison or a remedy, depending on how we care for it. In this chapter, I would like to illustrate how, in particular through the projects I have been able to carry out with him, Bernard Stiegler knew how to take care of what he called the "technologies of the spirit": first (1), by experimenting with them through cultural projects or technological development programs, whose incompletion he could criticize, thus building a critical vision and a *future* for technology; then (2), by taking care of them, that is to say, by considering them as objects of collective desire, but always pharmacological, in a necessary intermittence between what is synchronic and diachronic, between the stereotype and the traumatype,

Chapitre written by Vincent PUIG.

1 In his preface to the reissue of volumes 1–3 of *La technique et le temps*, (Fayard, 2018), Bernard Stiegler announces four new volumes: *L'épreuve de la vérité dans l'ère post-véridique* (4), *Symboles et diaboles* (5), *La guerre des esprits* (6), *Le défaut qu'il faut. Idioma, idios, idiocy* (7).

between what is calculable and incalculable; and finally (3), under the assumption that all knowledge proceeds from a transindividuation experience, by not even hesitating to consider this fundamental intermittence in the economic field of our "automatic society" through the development of the contributive economy project, between *ponos* ("labor") and *ergon* ("task"), between *negotium* and *otium*, between employment and work.

7.1. Memories and writings, retention and protention: constructing the organology of the spirit

Bernard Stiegler was a thinker of a general organology, which he derived from his reading of Canguilhem, Leroi-Gourhan and Simondon. This "taking care" was based on a remarkable capacity to desire and understand exactly what the organs (biological, technological, social or political) of this desire were, their conditions of possibility and their pharmacology. At the Institut de recherche et d'innovation (IRI), he knew how to develop these aspects down to the smallest details of technological or methodological implementation. A few weeks before he passed away, we had a lively discussion about the organology of the Zoom videoconferencing system, not just on the political aspects linked to its data extraction model but mainly on questions of attention design. In fact, he advised against using chat during seminars, rightly arguing that it would lead to a loss of attention, and finally pointing out that messages should not be displayed on the screen, but that the chat could be used by some participants as an annotation and indexing tool to supplement the video recording with keywords, bibliographical references, comments, etc. In a bold outburst in 2020, in front of representatives of the Thunberg generation, some of whom were born in the 21st century, he reactivated all the organological knowledge he had acquired since the 1980s.

With the *Mémoires du futur* exhibition (1987, Figure 7.1), which he was the curator of, Bernard Stiegler introduced the general public to the history and future of what he then called hypomnesic media and later named tertiary retentions and exosomatic organs: the emergence of writing, the first libraries, printing, postal networks, the telegraph, telephone, radio and television, as well as optical disks, editing benches, video recorders, telex, news writing tools, search engines and so on. Shortly afterwards, in 1989, as a precursor to the BNF, he contributed to the *poste de lecture assistée par ordinateur* (PLAO, computer-assisted reading) research group with writers, poets, musicians, scientists and computer scientists. There, he formalized a whole annotation device used for his thesis, notably with texts by Husserl, which was first modeled at the UTC and then implemented in 1992 on the SUN computer by Berger-Levrault (which later became Diadeis), under the direction of François Chahuneau. Much of this original resource has been lost under the weight of technological obsolescence and the increasing entropy of digital technology, built on a notion of information that is itself based on the implementation of the second law of thermodynamics which, to date, has found no alternative.

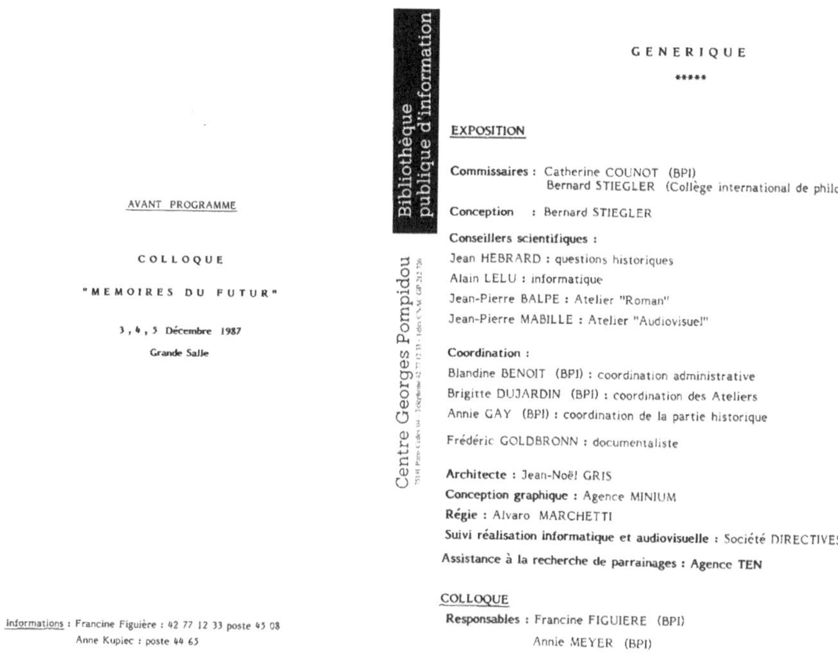

AVANT PROGRAMME

COLLOQUE

"MEMOIRES DU FUTUR"

3, 4, 5 Décembre 1987

Grande Salle

Informations : Francine Figuière : 42 77 12 33 poste 45 08
Anne Kupiec : poste 44 65

Bibliothèque publique d'information

Centre Georges Pompidou

GENERIQUE

*****

EXPOSITION

Commissaires : Catherine COUNOT (BPI)
Bernard STIEGLER (Collège international de philosophie)

Conception : Bernard STIEGLER

Conseillers scientifiques :
Jean HEBRARD : questions historiques
Alain LELU : informatique
Jean-Pierre BALPE : Atelier "Roman"
Jean-Pierre MABILLE : Atelier "Audiovisuel"

Coordination :
Blandine BENOIT (BPI) : coordination administrative
Brigitte DUJARDIN (BPI) : coordination des Ateliers
Annie GAY (BPI) : coordination de la partie historique

Frédéric GOLDBRONN : documentaliste

Architecte : Jean-Noël GRIS
Conception graphique : Agence MINIUM
Régie : Alvaro MARCHETTI
Suivi réalisation informatique et audiovisuelle : Société DIRECTIVES
Assistance à la recherche de parrainages : Agence TEN

COLLOQUE
Responsables : Francine FIGUIERE (BPI)
Annie MEYER (BPI)

Figure 7.1. *Brochure for the* Mémoires du future *exhibition, Centre Pompidou, December 1987*

In a text published in 1991 (Stiegler 1991), this *grammaticalization* protocol (Auroux 1995) is described according to four types of operation:

– *hierarchization* (creation of units of meaning by underlining or lines in the margin);

– *qualification* (keywords, annotations, comments);

– *navigation and search* (search windows, creation of links, link categories specifying the nature of attached documents: canonical commentaries, translations, manuscripts, references used by the text, commented on etc.);

– *representation* (various graphs, corpus "views", etc.).

In this system, there was no a priori categorization of possible annotation systems: the categorization tools were largely modular, following the logic of the SGML standard, which was HTML's predecessor. The concern for genericity and normativity, which is the current concern of W3C Open Annotation[2], for example, in its relationship with the RDF standard, is always at odds with the purely computational approach supported by statistical data analysis (automated categorization that does not need to be named, interpreted or standardized in terms of content). This is why the IRI has always opted for categorization systems that can be configured by the members of a workgroup and adjusted according to the steps the group sets itself[3].

Bernard Stiegler's constant encounter with these organological considerations through his philosophical thinking is similar to the way Simondon developed his theory of individuation from his mechanical and electrical tinkering. In Bernard Stiegler's case, however, technology was clearly presented as the foundation of what he called transindividuation following Simondon.

This encounter was then enriched in the 1990s at the CosTech Laboratory of the UTC with the "Territoires numériques" project[4], introducing the innovative concept of pre-indexed video forms, then at the INA, where he was deputy director general, and at IRCAM from 2001 to 2006, where I was able to create a hypermedia studio with him based on the initial model he had started at the INA with Jean-Pierre Mabille and Xavier Lemarchand. With the help of Nicolas Donin, this was where the *écoutes signes* ("signed listening") was developed (Figure 7.2) for instrumented listening and music annotation, which was later adapted at the Centre Pompidou and the IRI as the *Lignes de Temps* software (Figure 7.3), implemented in real time with Semantic HIFI for schools with MusiqueLab Annotation and promoted to

2 Available at: https://en.wikipedia.org/wiki/Web_annotation.

3 This was particularly the case following the SocialWeb project carried out at the IRI by Yuk Hui and Harry Halpin in 2012.

4 In partnership with Olivier Landau at Sofrecom.

amateur communities through the IRCAM Forum and the Résonance music technology festival.

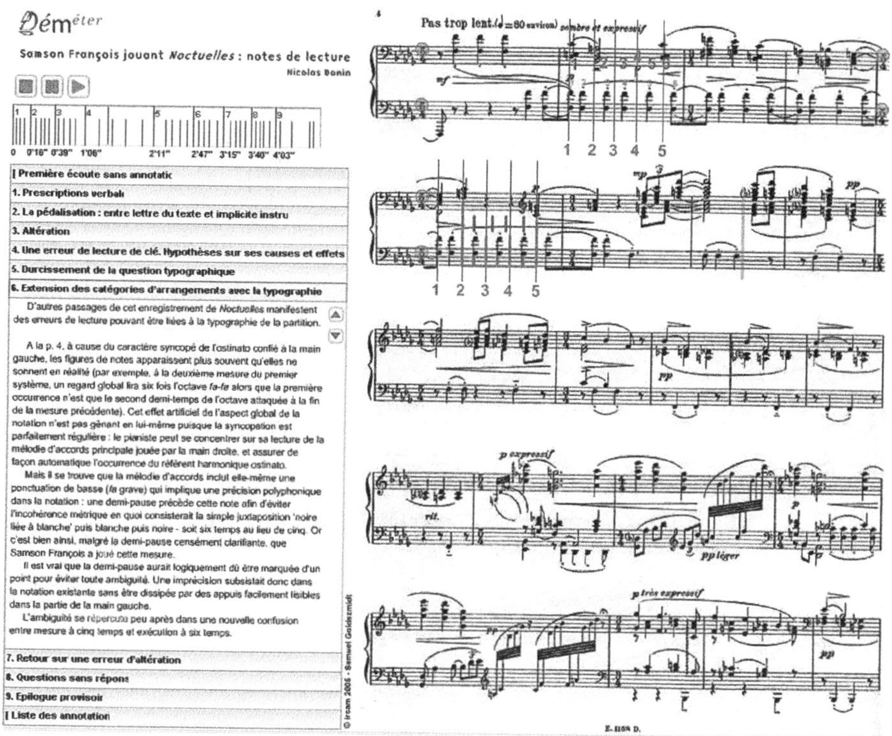

Figure 7.2. Écoutes signées *(signed listening) – IRCAM*

Whether they are text, music or film, all these technologies of the spirit are based on new forms of grammaticalization, formulating units of meaning, markings and modes of representation and navigation, as described in the 1991 article above. Bernard Stiegler had an in-depth knowledge of documentary techniques, but this was not limited to archiving or what came to be known as the digital humanities. In digital grammaticalization, he was looking for new knowledge-producing transductions (in Simondon's meaning of the term), innovative forms of knowledge-producing organization (negentropy).

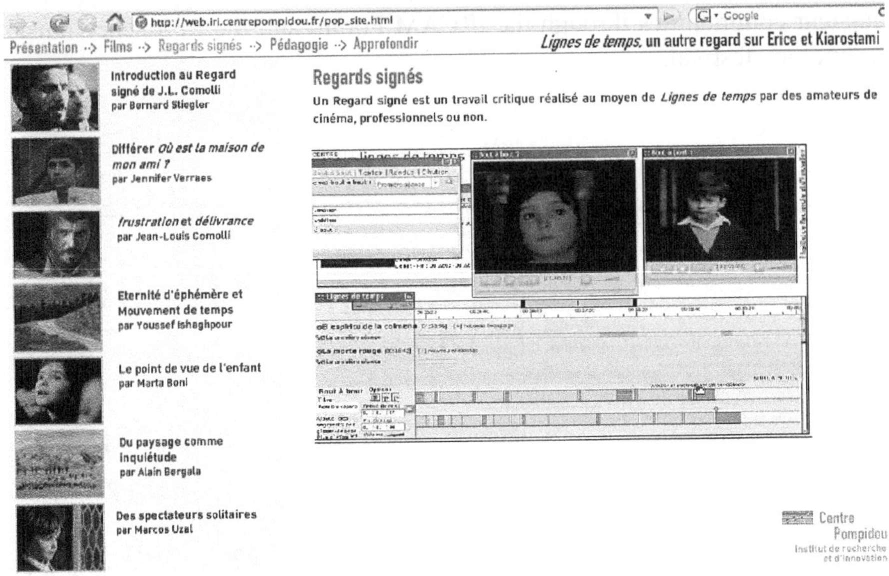

Figure 7.3. Regards signés *(Marked Looks) built with* Lignes de temps *– IRI, Centre Pompidou*

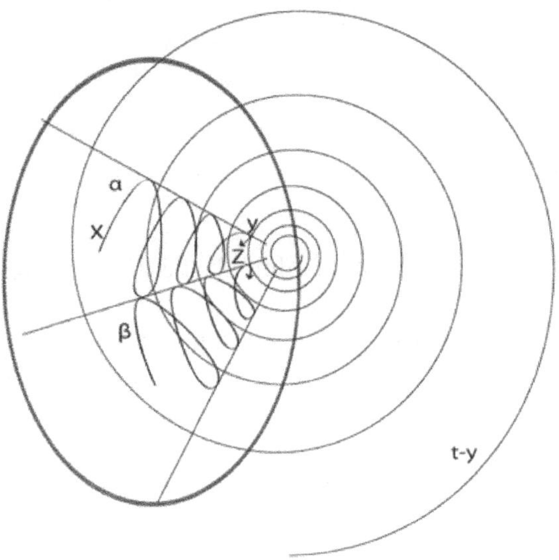

Figure 7.4. *Idiotext and spirals of individuation – Bernard Stiegler*

In a 2010 article, Bernard Stiegler shows how the *idiotext* (Figure 7.4), in other words, the unique process of writing, represented here by spirals of individuation, is always part of a retention and protention dynamic of memory and expectation that generates attention (Stiegler 2010). This idiotext is "a memory moved by its textuality", and it is built on "an irreducible difference that is technological". If technology is to be considered "reflexive", this difference, this founding "defect", must be constantly maintained. In many cases, Bernard Stiegler considered that this reflexive criterion was not maintained in a satisfactory manner, requiring frequent re-development or abandonment and, above all, constantly having to link the tool to a practice deemed contributive and negentropic. In terms of design, this reflexivity, which organizes time and space differently from entropy, is often maintained by transduction processes between reading and writing (which is why in *Lignes de temps* a simple full-screen reading must not be separated from annotation), between analysis and synthesis (which justified linking *Lignes de temps* with mental maps, for example), or between spatialization processes (segmentation of a video) and temporalization processes (oralization of a text).

7.2. Reflexivity for transindividuation

However, the organological foundations of this reflexivity design that Bernard Stiegler's thought calls for rest on a profound understanding of libidinal dynamics (Stiegler 2017) that always link, as it were, an automatic drive with an infinite desire, absolute synchronization to a video stream with a diachronic writing space, or phases of calculable hypomnesic understanding with phases of anamnesic disruption that are incalculable. This link, which Bernard Stiegler stipulates from his reading of Freud "between the dead and the living", is also a game of entropy retention thanks to temporal and memorial/historical processes that could, today, be described as anti-entropic (Longo and Montévil 2012) and which have recently been documented and commented on (Alombert and Krzykawski 2021).

One example of experimentation with this link was the development of contributory categorization protocols that promote *transindividuation*, first in a note-taking system for classes or lectures developed in the *Lignes de temps* software[5] in 2009, then in 2010 in the Polemictweet application[6], built on the Twitter API (Figure 7.5).

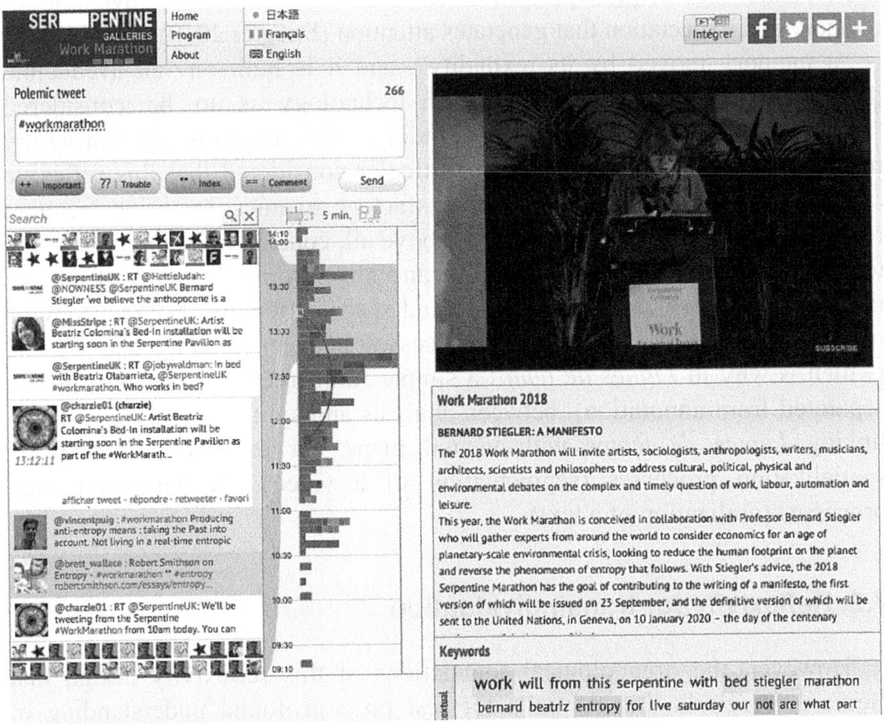

Figure 7.5. *Four meta-categories in Polemictweet – IRI. For a color version of this figure, see www.iste.co.uk/charolles/philosophies.zip*

The system works in three stages: (1) a presentation of the protocol to explain to contributors that their notes, which in this case are their tweets, will be published in sync with the course recording; (2) use of the live participation interface, which offers the insertion of four color-coded tags in tweets: comprehension (stereotypical retention marked in blue), disruption (traumatypical retention marked in yellow), keywords (marked in purple) and free comments (marked in green); (3) publication of the recording

5 Available at: http://ldt.iri.centrepompidou.fr.

6 Available at: http://polemictweet.com.

indexed by the tweets, allowing direct consultation of transindividuation points, keywords or references, as well as the use of an intra-video search engine based on the content of the tweets. These are meta-categories, since they do not directly relate to the content being annotated but to the meaning of the contribution. Subsequently, simply observing the co-occurrence of these meta-categories at any point in the course (Figure 7.6) creates a potential for exchanges between students.

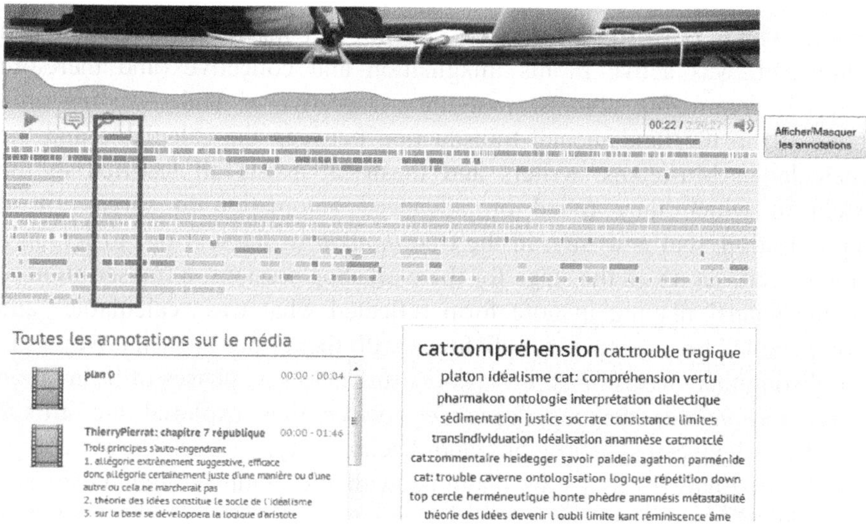

Figure 7.6. *Pharmakon course – vertical reading of areas of disruption (in red) on student-generated lines. For a color version of this figure, see www.iste.co.uk/charolles/philosophies.zip*

7.3. Taking care of intermittence

The materialization of this noetic intermittence between synchronicity and diachronicity, as Bernard Stiegler describes, does not lead to a formal, stable design. On the contrary, it needs to be "metastable" in Simondon's meaning of the term, in other words, always open to bifurcation and negentropic organization. These kinds of digital systems are not entirely diachronic, but play on synchronicity and the *understanding* with diachronicity, confusion and surprise – what Bernard Stiegler calls *surprehension*.

To illustrate this exploration and experimentation with noetic intermittence, we often present the *Penser-improviser* project (2015)[7], for it is undoubtedly in the context of art that Bernard Stiegler's thought opens up the most profound bifurcations. As part of the ceremonies of Mons, European Capital of Culture, at the Fabrique de Théâtre directed by Valérie Cordy, which was followed by Les Inattendus festival, there was a one-year rotation of musical improvisation workshops led by Bernard Lubat with talks from musicologists, writers and philosophers. The genuine function of reason, which Bernard Stiegler grasped through his reading of Kant and Whitehead, was active in his imagination and collective (and therefore spiritual) dream, but always from the perspective of understanding and, above all, in the fundamental transduction between the calculable and the incalculable. Musicians, writers and theatrical artists all testified to the extent to which they relied on advanced automation to surpass it in improvisation, that is to say, in the true *thought* of the text or music. On a large screen, set up on the stage for the workshop sessions, a transcription of the musicians' playing in note form reflected what was "calculable" and "computable" by a computer in dialogue with them[8], but in such a way as to also distinguish phases of improvisation, that is to say, phases of bifurcation in relation to this device. *Penser-improviser* thus explored the artistic boundaries between synchronicity and diachronicity, between the stereotype and the traumatype, in a *live* dialogue with the computerized music and in a "delay"[9] with the help of the Polemictweet system, used to annotate the musical categories emerging from the improvisation sessions, the performance sessions and Bernard Lubat's particularly colorful vocabulary. The annotations thus produced by the participants served as markers to facilitate replays on stage or on the *Lignes de temps* website (Figure 7.7).

This organology of the *amatorat*, with its multiple tertiary retentions and protentions, requires us to regain control over categorization, which, today, is largely carried out in a purely calculating manner by machines and through the production of statistical clusters of data.

7 Gaétan Robillard's video: https://www.ingenieur-imac.fr/realisations/penser-improviser; website designed by Simon Lincelles: http://penserimproviser.org.

8 ImproteK device by Marc Chemillier, musicologist at EHESS.

9 In computer music, this is known as "delay processing". See: https://docs.cycling74.com/max7/tutorials/15_delaychapter01.

Figure 7.7. Lignes de temps *for* Penser-improviser – *IRI (video)*

7.4. Toward a benevolent disposition

In the "Clinique contributive" workshop, which Bernard Stiegler started in 2017 in Seine-Saint-Denis with child psychiatrist Marie-Claude Bossière and researchers Maël Montévil and Anne Alombert[10], the aim, as part of a contributive research group (Stiegler 2020) involving parents, caregivers and academic researchers, was to collectively take care of the phenomenon of overexposure to screens in young children. This work, inspired by the institutional psychotherapy introduced by Tosquelles and the experience of Alcoholics Anonymous, as documented by G. Bateson (1971), analyzes digital socio-technical devices in their pharmacological aspect, while above all considering their potential for openness and desire, as well as their function as "potential and transitional spaces" (Winnicott 1971). In his thesis, taken up and interpreted by Louise Merzeau, and following Winnicott and Speamann (1997), Emmanuel Belin introduces the notion of a *benevolent disposition* (Belin 2002) considered as an "experience" which "allows one to see the day in a friendly light" and which is built on absence and defect forming a basis of trust, an "openness and recognition of the other", where

10 The "Territoire apprenant contributif" program project. See: https://tac93.fr/.

artifacts and human relations "co-construct in a movement of mutual reinforcement". Belin gives the example of a house and "inhabiting", as found in Bachelard, Blanchot and Heidegger. We find this concept interesting in relation to Bernard Stiegler's thought on defects and *filia* as conditions of possibility for knowledge. Even if Belin warns us here against any hasty implementation of benevolent disposition in the theological field – a field in which Bernard Stiegler knew how to distinguish the *pharmakon* with surprising knowledge – let us note that in the Catholic tradition, benevolence is not a (substantial) gift of the spirit, but must instead be sought in one's (existential) application: first and foremost, in *benignity* (goodwill, tolerance), but also in *mansuetude* (gentleness of the soul). However, in Gilbert Simondon's different, but not unrelated, concept of the spirit, the transindividual stage refers to a technical, but also spiritual, relationship between the individual and the group, and between the group and the individual, in a transductive relation between affectivity and emotionality. For Simondon, affective functions are oriented according to bipolarities that are meta-stabilized in emotions: affections (Simondon 2007), such as cheerful/sad, happy/unhappy, exciting/depressing, demeaning/enlightening, have a meaning revealed through emotion (visible from the collective), just like sensations linked to perception (hot/cold, light/dark) have a meaning revealed through action (anchored in the collective). In the digital studies we are developing at the IRI on the relation of calculable and incalculable functions, all knowledge production similarly relies on trust in the technical medium of writing, publication and, consequently, "transindividuation". Digital studies are therefore concerned with categorizing the conditions of knowledge production, without shifting into the production of ontologies of relations (in other words, into a potentially generalized calculability, like Facebook) or into emotional computing. In the Polemictweet system described above (Figure 7.5), the aim is to delay the synchronous real-time process of tweet production by using a tweet categorization protocol that allows diachronic uses (formation of consensus or dissensus groups) during the event, and especially afterwards. The system does not, therefore, simply highlight emotions, or even affections, which are vectors, functions and "indications of becoming" (Simondon 2007). Instead, it responds to Simondon's objective of moving from emotion to perception and from perception to emotion. It also aims to articulate affection and action, the individual aspect of psychic becoming and the collective aspect of knowledge production.

In a "benevolent" social network, that is to say, one designed for interpretation, not only do individuals need to be able to confidently project themselves, but groups need to be able to make collective decisions, in other words, deliberate decisions, and projects need to be able to be editorialized by these groups. Discussions also need to be graded according to the level of collaboration. For example, exchange and sharing functionalities must be accompanied by the possibility for group members to compare their work, and thus annotate each other's annotations. Yuk Hui and Harry Halpin have shown that a social network like Facebook is built on the principle of Moreno's social graphs, which is the idea that the individual is the primary node in the network. In addition to this techno-methodological individualism, which tends to favor the notion of a profile that no longer produces individuals, but "dividuals" instead (Deleuze 1990), Yuk Hui and Harry Halpin (2013) have proposed an approach to social relations based on the group, in Simondon's sense. The latter tends to value the *associated milieu*[11] and therefore formalized sharing functionalities between network members. The individual therefore no longer comes first, and what is favored is their relationship to the associated environment: their affiliation to one or more groups, their work on one or more projects, etc. This involves having a different relationship to computation. Rather than seeing algorithms as a means of *extracting* data from a user, in order to predict his or her behavior, group-based social networks must rely on algorithms for recommending hermeneutic convergences or divergences, in other words, the automated suggestion of groups to be formed. Such an approach also implies offering network user-contributors greater leeway in terms of network governance and development.

Indeed, autonomy is what initially characterizes a group, in other words, the latter's ability to set rules for itself, or normativity in Canguilhem's sense (1966): a hermeneutic group must be able to decide how to organize the sharing of contributions, including in terms of network architecture (multicast and decentralized) and data (open to deliberation). This can mean the group questioning of the meta-categories that make up its common annotation language, or deciding whether to publish a collective work. However, it also means that the group can transform its workspace, by contributing to the open co-design of the categorization platform. All this

11 For Simondon, individuation proceeds from a transductive relationship between the pre-individual's potential and his associated environment, in a non-ontological surpassing of von Uexküll's *umwelt*.

implies that the social network must be equipped with a governance system such as the one formalized in Wikipedia, even if it is gradually leaning toward automation. This network design that is fundamentally attached to a work dynamic – collaborative, because it is based on joint work, and contributive, because it is dedicated to a form of publication that recognizes contributor input – must use calculation for deliberative purposes and not directly for prediction or performance.

7.5. The practice of knowledge and the contribution economy

With the launch of the "Territoire apprenant contributif" program, Bernard Stiegler and the IRI team have progressively developed the model of the contribution economy based on a fundamental noetic intermittence between automations and bifurcations. In concrete terms, this intermittence happens between periods of knowledge development supported by a contributory income, and periods of implementing this knowledge in the economy, thereby renewing income rights. This model is inspired by another historical form of intermittence, that of the compensation scheme for intermittent workers in the entertainment industry (in France and Belgium), but against a backdrop of declining employment due to automation, which requires us to think of other forms of commons production, such as those found in the world of free software (Morlat et al. 2021). This intermittence is based on a very specific difference and articulation between employment and work, in other words, between labor (*ponos*) and work (*ergon*), between *negotium* and *otium*, or the French *labeur* and *œuvre*, respectively. Here, following André Gorz, Antonella Corsani (2020) distinguishes between heteronomy and autonomy.

At this time when the health crisis is blurring all the boundaries between the private world and the world of work, we urgently need to think about the intermittence of our minds and actions, between our automations and our bifurcations, between the calculable and the incalculable and between desire and drive. The Covid-19 crisis manifested and increased all kinds of "divides", it is a new issue whose pathological effects must still be measured if we are to even start thinking of intermittence as a valid option against entropic synchronicity, or as Antoinette Rouvroy put it, to develop *heterochronicity* (Rouvroy 2015).

In all the "Territoire apprenant contributif" workshops[12], a constant concern to encourage negentropic bifurcations can be seen through the method of analyzing the value of the practice of knowledge, enabling it to be accounted for and supported by a contributory income or a local currency. It is here, at the heart of a system for empowerment and the development of economic activities, that the organological foundations introduced by Bernard Stiegler are continuing to unfold.

7.6. References

Alombert, A. and Krzykawski, M. (2021). Vocabulaire de l'internation. *Appareil* [Online]. Available at: http://journals.openedition.org/appareil/3752.

Auroux, S. (1995). *Les révolutions technologiques de la grammatisation*. Mardaga, Brussels.

Bateson, G. (1971). The cybernetics of self: A theory of alcoholism. *Psychiatry*, 34, 1–18.

Belin, E. (2002). *Sociologie des espaces potentiels. Logique dispositive et expérience ordinaire*. De Boeck, Brussels.

Canguilhem, G. (1966). *Le normal et le pathologique*. PUF, Paris.

Corsani, A. (2020). *Chemins de la liberté. Le travail entre hétéronomie et autonomie*. Éditions du Croquant, Vulaines-sur-Seine.

Deleuze, G. (1990). Post-scriptum sur les sociétés de contrôle. *L'Autre Journal*, 1.

Hui, Y. and Halpin, H. (2013). Collective individuation: The future of the social web. In *Unlike Us Reader: Social Media Monopolies and Their Alternatives*, Lovink, G. and Rasch, M. (eds). Institute of Network Cultures, Amsterdam.

Longo, G. and Montévil, M. (2012). The inert vs. the living state of matter: Extended criticality, time geometry, anti-entropy – An overview. *Frontiers in Physiology*, 3(39). doi: org/10.3389/fphys.2012.00039.

Morlat, C., Sentis, T., Landau, O., Kunvari, A., Puig, V. (2021). Économie de la contribution et gestion des biens communs. *Imaginaire Communs* [Online]. Available at: https://anis-catalyst.org/imaginaire-communs/economie-de-la-contribution-et-gestion-des-biens-communs/.

12 See: https://tac93.fr/.

Rouvroy, A. and Stiegler, B. (2015). Le régime de vérité numérique. *Socio* [Online]. Available at: http://journals.openedition.org/socio/1251.

Simondon, G. (2007). *L'individuation psychique et collective*. Aubier, Paris.

Spaemann, R. (1997). *Bonheur et bienveillance. Essai sur l'éthique*. PUF, Paris.

Stiegler, B. (1991). Annotation, navigation, édition électroniques : vers une géographie de la connaissance. *Linx, Texte et ordinateur. Les Mutations du Lire-Ecrire*, 4, 121–131 [Online]. Available at: https://www.persee.fr/doc/linx_0246-8743_1991_hos_4_1_1191.

Stiegler, B. (2010). Le concept d'idiotexte : esquisses. *Intellectica. Revue de l'Association pour la Recherche Cognitive*, 53/54, 1–2.

Stiegler, B. (ed.) (2014). *Digital Studies. Organologie des savoirs et technologies de la connaissance*. Fyp, Limoges.

Stiegler, B. (2017). Organologie de la libido. *Deleuziana*, 6, 68–81.

Stiegler, B. (ed.) (2020). *Bifurquer*. Les Liens qui Libèrent, Paris.

Winnicott, D.W. (1971). *Jeu et réalité*. Gallimard, Paris.

8

Predictive Machines and Overcoming Metaphysics

8.1. Cybernetic machines and intelligent machines

Cybernetic[1] machines were inspired by the self-regulatory mechanism that enables organisms to preserve their functional balance by responding to changes in the environment that they are adapted to. As a science, cybernetics aims to capture the information-processing procedures that characterize classes of systems and represent them as a sequence of actions (outputs) correlated to variations in the values of relevant parameters (inputs). New-generation intelligent machines, on the other hand, are influenced by the learning mechanism by which humans modify operating rules in a standardized way, forcing the environment to adapt to their needs. The evolution of simpler organic systems is brought about by externalities (the influx of unexpected information) and the natural selection of random changes that are appropriate to the new situation, whereas humans intentionally advance their behavior through inductive reasoning, the basis of scientific method: the selection of hypotheses that guide behavior is made on the basis of efficiency toward freely determined purposes. Cybernetic machines are therefore the result of scientific hypotheses about the information-processing rule that makes the behavior of biological systems (capable of performing tasks) predictable and exploitable, while intelligent

Chapter written by Anna LONGO.

1 Norbert Wiener defines cybernetics as "the study of control and communication in animals and machines" (Wiener 2014, p. 70).

machines are the result of scientific hypotheses about the learning formula by which humans progress in modeling physical systems. This representation of systems as information structures not only makes it possible to predict their behavior in order to interact in a productive manner – for example, by using them to get a job done – but to also build machines that optimize organism operations and take their place in carrying out tasks. In the same way that cybernetic machines automated the functions of organic bodies, improving their efficiency, AI aims to automate the production of knowledge, making it faster than the computing power of the biological brain. This project assumes the establishment of a science of learning processes, by which humans change hypotheses about the functioning of physical systems, as well as the behaviors resulting from their operation or technical reproduction. AI is thus the result of the introduction of an empirical science of knowledge, which, as an objective, considers the way in which a rational agent voluntarily modifies its practices and modes of interaction, going beyond the genetically inherited information-processing procedures it shares with other animals, which have a very slow and accidental evolution. Artificial intelligence systems can be seen as experiments designed to confirm the reasoning procedure hypothesis, which enables humans to discover new patterns in the systems with which they interact, and to select (culturally transmit) hypotheses about how humans relate to their environment in such a way as to guarantee the unlimited growth of the species and its material (or economic) satisfactions. Rather than simulating the functions of organic bodies, AI aims to simulate the cognitive procedure by which humans manage to evolve their interaction strategies through progress in their understanding of how physical systems work and thus through continual criticism of the scientific hypotheses that guide decisions. However, as we shall see, this also means that humans are no different in kind from other organic systems, since their behaviors can be explained, anticipated and optimized by grasping the information-processing procedure that characterizes them.

It is important to note that hypotheses about the cognitive process inherent in inductive learning are scientifically acceptable, unlike philosophical speculations about the essence of thought and the metaphysical principles that strive to make it relevant to reality. AI is therefore the result of the extension of scientific inquiry into the rational procedures of science. AI is the technology that should enable this inquiry to progress by providing research with an exceptional capacity for information processing. The realization of a more perfect AI is correlated with an understanding of the cognitive operations by which humans elaborate models that are increasingly in tune with the information-processing rules that characterize organic systems and by which they realize AI. Furthermore, it is because of the computing power implemented by this technology that the complex formula of knowledge production will finally be clarified[2].

By taking up the contrast between Carnap and Heidegger, we would like to show that AI, conceived as a technology for automating the production of knowledge, presents a real challenge to philosophy in its traditional role as the "science of knowledge". If the philosophical question concerns the condition of scientific knowledge that science cannot investigate, what happens to philosophy once science becomes capable of grasping the cognitive process that makes it possible?

8.2. The overcoming of metaphysics and the automation of knowledge production

In order to better understand the challenge that artificial intelligence presents to philosophy, it is important to consider that the automation of knowledge production is part of an effort to establish an objective method of assessing scientific hypotheses, which characterizes the anti-metaphysical stance of logical empiricism. As we can clearly see from Carnap's project, the aim is not only to assess the level of confirmation (probability) of inferences or predictions based on repeated experimental observations (data) but also to construct a logic of induction, in other words, to clarify the system of rules that makes the progress of knowledge possible. Consequently, the problem of induction is reformulated in terms of the rational criteria, which enable the selection of the most reliable hypotheses, that is to say, hypotheses that everyone should consider as the most likely for making optimal decisions. By logically analyzing the language of scientific theories, we should be able to isolate their atomic terms and then determine the consequences, in other words, the predictive suggestions objectively involved. The aim is to reconstruct the logic of inductive reasoning from its historical productions, that is to say, from the syntactic and semantic analysis of historically selected and regularly used hypotheses. This means clarifying

2 The idea that AI will shape the future evolution of human organizations is very explicitly expressed by theorists of the technological singularity (Kurzweil 2007).

the rules of inductive learning, so as to facilitate the progress of knowledge by tracking down any beliefs that do not conform to the axiomatic definitions that found the building of objective knowledge, or even the set of logically acceptable and empirically verifiable propositions. The consequences for philosophy are well-known. As Carnap states, logical analysis becomes the essential method of reflection for freeing the field of knowledge from the false beliefs and illusions of metaphysics (Carnap 1985). Philosophy thus finds its true role in the logical analysis of scientific language and finds its purpose in the selection of hypotheses that are in line with the rational norm of induction. It is this function of controlled criticism that can potentially be automated: once the axiomatics of inductive reasoning have been established on the basis of the analysis of empirical cases (the scientific propositions universally held to be true), all that needs to be done is to mechanically verify whether any new hypothesis proposed is acceptable, and if it can be produced from atomic terms. The recursive functions used by Turing machines enable such a calculation to be carried out, although it is necessary to understand the scientific hypotheses as rules capable of generating sequences of values that represent the variation curve of a phenomenon, rather than as inferences logically implied by premises. Computers can actually be used to check whether a certain event is part of a series of values produced by a certain function, and thus whether a certain prediction conforms to a hypothesis that is considered as reliable. Furthermore, the reliability of hypotheses can be verified by calculating whether an observed fact was actually foreseeable, that is to say, whether this fact was part of the sequence generated by a function or hypothesis. This makes it possible to assess competing predictive hypotheses by calculating their respective probabilities, as well as to select the most suitable hypotheses for regular use (in other words, those to be assumed as true).

Ray Solomonoff, the father of artificial intelligence and a student of Carnap in Chicago, was the first to understand this, as well as the first to use a Turing machine to automate inductive reasoning. However, he also understood that to do so, it was necessary to abandon the logical framework and embrace the Bayesian concept of knowledge production. Instead of measuring the level of confirmation of theories on the basis of the logical implication between premises (data) and consequences (predictions), Solomonoff suggested comparing the different predictive hypotheses that can be derived from available data by applying Bayes' formula. This ancient theorem – revised in the 1950s by Leonard Savage to produce a rational agent model (Savage 1954) – enables us to objectively calculate how new observations modify the probability that was initially attributed to a set of predictive hypotheses; this is the basis of today's machine learning. To automate inductive reasoning, Solomonoff considered that any predictive hypothesis can be expressed as a recursive function, capable of generating the complete series of variations by which a phenomenon manifests itself (e.g. the function that produces the viral infection curve, making it possible to predict the evolution of the phenomenon from the analysis of available data at a given moment). Each phenomenon is thus represented as a pattern of information that can be produced by an algorithm. Bayes' theorem is used to select the most probable algorithms, those whose predictions are the most reliable at a given point in time (subsequent observations may modify the respective probability of plausible hypotheses).

Without going into the details of the theory that allows Solomonoff[3] (1997) to lay the foundations of contemporary artificial intelligence, it is important to note that the project extends the Carnapian endeavor to grasp the objective rules of inductive learning, so as to optimize the process of knowledge evolution. Solomonoff's inductive machine is, in fact, capable of finding a large number of patterns in the data analyzed and, consequently, of producing a large number of hypotheses or functions that are capable of generating the observed values in the form of ordered sequences. Moreover, it is not only capable of determining the respective probability of these hypotheses (according to the principle of algorithmic probability[4]), but also of changing this judgment by considering the addition of new information. In short, Solomonoff's intelligent machine prototype is capable of replacing scientists in the production of plausible hypotheses (acceptable in relation to the available information), and philosophers in the assessment of acceptable propositions – those that everyone should assume as true in order to make effective decisions.

New intelligent machines therefore aim to simulate the inductive reasoning that enables humans to evolve their knowledge, by means of a continuous process of re-assessing acquired knowledge in light of new

3 Solomonoff was one of the first to produce artificial neural networks capable of inductive learning.

4 Algorithmic probability measures the ex ante probability that must be attributed to hypotheses that have been identified by exploring a database in relation to the complexity of the algorithms likely to generate the desired series. According to Solomonoff, simpler or even shorter rules are a priori more probable.

observations. The human being is thus potentially reduced to a computational procedure that cannot only be simulated, but can also be technically optimized. Just as a physical or biological system is objectively represented as a sequence of information (outputs) that can be generated by a more or less complex algorithm, it would seem that human learning can be reduced to a formula for automating and accelerating the production of knowledge, as well as the productive evolution of practices. Although we do not yet have this formula, and today's artificial intelligence is still just a means of testing various hypotheses on the computational procedure of knowledge, it must be acknowledged that the successes already achieved seem to confirm the idea that the human being is objectively knowable and can be identified with a pattern of information, which only differs from other physical systems in the degree of complexity of the algorithmic rule, enabling its reasoning to be correctly simulated.

As Heidegger predicted, the danger involved in the implementation of technology is the human being's reduction to an object of scientific knowledge, or even an exploitable resource (Heidegger 1985, p. 36), like all other natural entities. Having become the privileged objects of scientific research, humans are the most valuable source of information for technical progress. Human agents are objectively known for their decision-making patterns, which must be predicted in order to administer the inputs needed to obtain the desired output (buying a product, reacting on social media, voting for a certain candidate, etc.). However, at the same time, they are constantly providing new observations, such as reactions that cannot yet be predicted or simulated. These are immediately used to update the predictive hypotheses that AI relies on and to possibly optimize the productive efficiency of new technologies.

The dispute between Carnap and Heidegger, who saw a danger from the result of reducing the activity of thinking to a formalized and automated procedure, therefore remains relevant. In "The Elimination of Metaphysics Through Logical Analysis of Languages", Carnap (1985) cites numerous expressions from *What Is Metaphysics?* (Heidegger 1990) as examples of propositions that are meaningless, in other words, that do not conform to the logic of induction, and must be discarded. According to Heidegger, the role of philosophy is to question the conditions of the scientific representation of being, a question that science is not in a position to ask. Carnap's response is that, on the contrary, the conditions of knowledge can be investigated in accordance with the requirements of scientific truth: one only needs to consider inductive reasoning as a procedure that obeys a certain number of rules, to which philosophy would itself be subject in its role of verifying the acceptability of statements. As Heidegger understands it in "The Question Concerning Technology", the danger of reducing the human to a mere resource is implied by the more serious danger of eliminating "metaphysics", by making philosophy an activity at the service of technological unfolding. In other words, it is by depriving thought of the freedom to question the necessity of its determination, by the current norm of scientific induction[5] (which provides the criteria for what is true and what is false in relation to the description of empirical entities), that the human is entering this dangerous path of complete enslavement to technology, making itself available as a resource to be exploited. It seems to me that the automation of knowledge production, like the industrial production of information, has already been denounced by Heidegger when he criticized Carnap's project, which reduces philosophy to an activity driven by the imperative of productive efficiency, or even permanent technological innovation. Today, it seems that new intelligent machines are evolving in an unpredictable way, feeding off all our thoughts, decisions and criticisms to offer us renewed solutions. Paradoxically, this increased freedom comes from the fact that we are constantly assisted in our decisions by portable computing devices, capable of guiding our choices between a growing range of options that are tailored to our future identities, beliefs and political commitments. However, we cannot help wondering if there is anything we can say, do or think that does not fuel the automated information production industry on which innovation and growth depend, as well as the unequal distribution of benefits and the depletion of the ecosystem. The timid persistence of this apparently unanswered question carries the last residue of philosophical freedom, and gives us hope for a new beginning in thought, not a step forward in knowledge in line with current norms, but a genuine *poiêsis*: the creation of a technology that enables us to see the world as something other than a set of information patterns.

---

5 As Heidegger states: "Where Enframing holds sway, regulating and securing of the standing-reserve mark all revealing. They no longer even let their own fundamental characteristic appear, namely, this revealing as such" (Heidegger 1958, p. 37).

8.3. References

Carnap, R. (1985). L'élimination de la métaphysique par l'analyse logique du langage. In *Manifeste du Cercle de Vienne et autres écrits*, Soulez, A. (ed.). PUF, Paris.

Heidegger, M. (1958). *The Question Concerning Technology and Other Essays*. Lovitt, W. (trans.). Harper and Row Publishers, New York.

Heidegger, M. (1990). *Questions I et II*. Axelos, K. (trans.). Gallimard, Paris.

Kurzweil, R. (2007). *Humanité 2.0 : la bible du changement*. M21, Paris.

Savage, L. (1954). *The Foundations of Statistics*. Dover Publications, New York.

Solomonoff, R. (1997). The discovery of algorithmic probability. *Journal of Computer and System Sciences*, 55(1), 73–88 [Online]. Available at: http://raysolomonoff.com/publications/barc97.pdf.

Wiener, N. (2014). *La cybernétique : information et régulation dans le vivant et la machine*. Le Seuil, Paris.

9

Artificial Intelligence's New Clothes

In 1770, Hungarian inventor Johann Wolfgang von Kempelen exhibited his automaton chess player, the "Mechanical Turk", for the first time at the Viennese court of Empress Maria Theresa. Seated at a wooden console, the automaton was dressed "in a Turkish style". The cabinet's side doors opened onto a set of cogs and gears that supposedly operated the automaton chess player's body. The Mechanical Turk would go on to win numerous victories – its most illustrious would be against Napoleon Bonaparte and Benjamin Franklin. Not only could it solve the tricky mathematical-logical problem of the knight, but it would also have been able to recognize cheating or an illegal move by its opponent. For over 80 years, the automaton toured Europe and the United States, meeting dignitaries, inventors and great chess players, before being destroyed in a fire in 1854.

The Mechanical Turk brings about two deeper issues that we would like to highlight and explore here, in the context of AI's contemporary resurgence: (1) mechanism or automatism, as a blind sequence of causes and effects, is always the automation of the other; (2) intelligence refuses to let such an order out of its sight and control, as it experiences its power and enjoyment by exercising this control. These two issues come together in what science historian Simon Schaffer (1994) describes as the "geography of intelligence", which assumes the spatial distribution and communication of human and machine behavior.

Chapter written by Tyler REIGELUTH.

9.1. The automation of the other

From the 18th and 19th centuries, the design and advertising of automata were influenced by forms of orientalism, colonialism and sexism. "The neat connection between passion, exoticism, mechanism and money permeated the showrooms. [...] The slippery move between images of languorous oriental baubles and honest utilitarian labor defines the significantly ambiguous place the automata occupied in a metropolis equally impressed by the mechanical ingenuity, excessive wealth and eroticized luxury which all marked its new world-wide imperium. This was an apt stage for automatic Turks, mechanical elephants, and clockwork women" (Schaffer 2016). The history of automata is part of a network of socioeconomic expectations and production: the tensions between artisanal and industrial work; the secrecy of production and the world's fair; the civilizing of manners and the staging of foreign cultures. The unfathomable yet spectacular nature of these automata is reminiscent of the state of our contemporary technical culture in the face of developments in artificial intelligence, the operations and structures of which bring us back to certain experiences of technical alienation (Reigeluth 2022).

As Schaffer points out, attributing intelligence to automata depended as much on concealing their internal mechanisms as it did on denying the work of human intelligence involved in their construction, in favor of the "inventor's" genius alone (Schaffer 2016). The automaton was the material proof of the "inventor's" genius, who alone possessed the secret of its operation. Access to knowledge of the machine necessarily required knowledge of the creator's plans and, after all, the automaton's mechanism was still nothing more than the execution of a programmed discourse. An algorithm, in the traditional sense of the term, is but a procedure for certifying a result, a set of perfectly ordered steps that cannot fail to produce a given result (Bachimont 2010). In this sense, an algorithm is first and foremost a control mechanism for ordering disparity toward an objective end. We know that the contemporary concealment of algorithmic logics, notably through commercial or industrial secrecy, is both the wealth of platforms and the effectiveness of new artificial intelligences (Cardon 2013). However, we also know to what extent this power depends on an extractive economy and an expendable workforce, which tends to disappear in the seamless design of devices and interfaces that users interact with (we will come back to this).

The story of automation is seldom one of the gradual displacement of human work, which is all too linear and easy. More often than not, it is the story of redistributing the geography of intelligence (Collins and Kusch 1998). The "mechanization" of a certain human activity always initially implies it being downgraded and socially divided as labor with no inherent intelligence, which can be governed from the towering gaze of the commander, the planner or the manager (Daston 1994; Canguilhem 2009). The searing scandal symbolized by the Mechanical Turk for writers such as Robert Willis and Edgar Allan Poe (1836), who tried to prove that appearance is not what makes the automaton, was that a machine could not only simulate the mechanical movements of the body, but also, and above all, the player's human intelligence. However, this distinction did not apply to just any human, but for those with a social status that made them invulnerable to the effects of automation, those who were now a part of the "majority". "Enlightened science imposed a division between subjects that could be automated and those reserved for reason. Such a contrast between instinctual mechanical labor and its rational analysis accompanied processes of subordination and rule. It seemed as if most subjects had never been, could perhaps never be, enlightened" (Schaffer 1999, p. 164).

The Mechanical Turk and the repeated attempts to reveal it marked an ambivalent high point in the Enlightenment project. The automaton seemed not only capable of executing rules, but also of making reasonable use of them, thus setting itself apart from the prowess of other automata of the Enlightenment era, whose mechanisms could be revealed to have been skillfully crafted by watchmakers, simply by opening their bodies. Something about the Mechanical Turk resisted the possibility of this exposure (and not just the orientalist costume), as nothing in its assembled mechanisms allowed us to *understand* how it adopted behaviors based on its opponent's behaviors. There was a *gap* between the causality of its mechanism and the effects it produced. The experience of this *gap* is the minimum condition for imputing intelligence to the observed behavior, since irreducibility to mechanical causality is grounds for a teleological argument (Kant 2000) that leads us to *act as if* the machine had a purpose of its own.

The Enlightenment's opposition to a mechanistic approach of anthropology and politics simultaneously produced a social divide between, on the one hand, those who, through cowardice or idleness, had surrendered to their first nature and, on the other hand, the enlightened who, through education, had acquired a second nature that conformed to reason. Historian Minsoo Kang (2011) reminds us that the mechanistic celebration of the Cartesian automaton in the 17th century gave way, during the 18th century, to a distrust for any human assimilation to the automaton, the latter being reserved for the mentally handicapped, the ignorant and the oppressed (p. 159) The automaton-human expressed a relationship of dependence and guardianship, an "inability to use one's understanding without guidance from another" (Kant 2008, p. 5).

However, far from putting mechanism to death, the Enlightenment paved the way for the latter to be implemented in the worker, the compliant laborer, whose function was to perform their task in the great production algorithm. In Kant's "What is Enlightenment?" (2008), we find an acceptable degree of mechanism at the community level, but only insofar as it serves, and is transcended by, a rational end: "[...] for many affairs conducted in the interest of the commonwealth a certain mechanism is necessary, by means of which some members of the commonwealth must behave merely passively, so as to be directed by the government, through an artful unanimity, to public ends (or at least prevented from destroying such ends)" (pp. 8–9). In other words, it is not a question of distinguishing between those who are now part of the majority and those who are still in the minority, but the *activity* of the reasonable majority that requires and assumes, on a collective scale, the *passivity* of the minority, on which it can exercise and prove its worth.

Even today, algorithmic learning systems, which automate the learning process and not just manual labor or information processing, rely heavily on low-skilled, fragmented and invisible labor. Amazon's Mechanical Turk platform offers companies expendable, collective and distributed workforce solutions to problems that automated computing power alone cannot solve. These tasks, which are beyond the reach of the machine, are called human intelligence tasks (HITs). However, behind this "intelligence" is a vast army of "click workers" who spend their hours clicking away under strenuous, flexible work regimes. Workers classify images, identify emotions or label hate speech, in order to build up training data for learning algorithms. They are like nannies or trainers for algorithms that are still minors, as we cannot hold them accountable for their actions and we currently do not have any educational theories that would enable us to say what machine learning is from a social point of view (Reigeluth and Castelle 2021). The "intelligence" of these workers that make machine learning possible is not that of any one individual, but an intelligence that is aggregated into a collective work force. While the consumer buys the product of this collective labor, the worker is paid pennies for his or her individual gestures; a marginal price for a marginal movement. "This is why the kind of algorithmic digital labor that is produced on Mturk is always invisible surplus labor and, to a certain extent, free, expropriated"[1] (Casilli and Cardon 2015, p. 92).

9.2. (Un)controlled intelligence

Although today's AIs are no longer dressed in a "Turkish style", they nevertheless convey similar fears, power relations and, above all, distributions of intelligence. Predictive algorithmic systems that grind data continuously and are coupled with user behavior exhibit a "margin of indeterminancy" (Simondon 2012, p. 193); the "algorithm" is less and less a control mechanism in the deterministic sense of the term. The fear of losing control over technology is a recurrent theme in Western thought on technology (Winner 1977). Responding to these fears by reasserting human control[2] ("it is always the human who decides", "you can always unplug the machine") seems to bring back a certain "master morality", to put it in Nietzschean terms, which bases any asymmetry of power on the fact that one dominates over the other. This does not entail defending the ontological position that the machine is truly intelligent or that it should enjoy moral autonomy. The point is simply to recognize that what we call "artificial intelligence" requires us to radically rethink the purpose of our technical knowledge and the legitimacy of its practice. When a Google image-recognition algorithm classifies black-skinned people as "gorillas" on the basis of training data, or a Microsoft chatbot makes negationist, racist and misogynistic remarks that it has learned from its social interactions (Lee 2016) – the exact reasons for this result being quite difficult for engineers to determine – this is not just a question of responsibility or ethics in relation to a technical problem but, more profoundly, a political issue in terms of who can grasp and value the "intelligence" of these systems.

---

1 Original French extract: "C'est pourquoi le type de *digital labor* algorithmique qui est produit sur Mturk est toujours un *surplus labor* invisible et, dans une certaine mesure, gratuit, exproprié" (Casilli and Cardon 2015, p. 92).

2 For a recent analysis of this topic, see the thematic issues "Contrôler l'intelligence artificielle?" of *Réseaux* (232/233).

In a 2018 song, *Blowing up the Congo*, the band Petite Noire and rapper Saul Williams encourage us to think about the relationship between colonialism and technology, and more specifically, between a colonial mentality and a fear of AI:

> What are you surprised? Master slave binary cobalt rubber refinery colonial mentality fed into machines, dissect the engines and devices, not just the precious stones that make them run, but the naming of the parts where one has uni-directional control over the others, one part of the engine is called the master, the other parts are called the slave, the word "robot" means slave, the fear that the slaves would learn too much and takeover is the fear that the robots would learn too much and takeover, takeover the minds of the Congo, South Africa, Mozambique, Zambia, Guinea, Nigeria, Senegal, Botswana, Angola, Ghana, Tanzania, Namibia, Mali, Egypt, Libya, Algeria… Oh what a surprise, fear of Africa and AI.

The song obviously refers to the disastrous extraction of minerals used in electronic components, both ecologically and economically. The new geography of intelligence that underpins AI production is a planetary multi-scalar process that ranges from extractive economies to distributed labor and data-capture structures (Crawford 2021). However, the song also, and perhaps above all, points the finger at the colonial mentality, a slave logic at the heart of digital devices, where everything functions according to chains of orders, commands and execution, and where the operation of any machine derives from the rules that govern its purpose (Eglash 2007). The song establishes a homology between the colonial system's fear of the colonized who might revolt and fear of an AI revolt. This is a fear that does not misrepresent the state of affairs linked to the actual capabilities of "intelligent" systems, but rather the prospect that their increasingly irreducible purposes will leave us with nothing on which to exercise our intelligence. Simondon (1958) made a similar diagnosis when criticizing the "yield/productive morality" that reduces technical performance to the maximization of abstract functions: "[…] despite appearances and despite the apparent civil liberties it allows individuals, a civilization of efficiency is extremely constraining for them and prevents their development, because it simultaneously enslaves man and machine; through the machine, it achieves restrictive communal integration. It is not against the machine that man, under the influence of a humanist concern, must rebel; man is only enslaved to the machine when the machine itself is already enslaved by the community"[3] (p. 355).

The revolt of the subalterns in the political context extends into the myth of the automaton-robot revolt. As Dumouchel and Damiano (2016) point out: "We want robots to have all the qualities that masters look for in their slaves, bosses in their employees, commanders in their soldiers, and we do not want them to have weaknesses, shortcomings or, least of all, that irrepressible tendency to insubordination, to independence, and to doing as they please, like that which characterizes human workers"[4] (p. 10). Now, in the era of machine learning, the question is how we can *reasonably* demand that a machine be both perfectly obedient and capable of learning.

However, beyond the question of whether a machine is really capable of learning or is endowed with intelligence, what should interest us here is the response provided, the solution demanded, in situations where the power we are used to exercising over machines seems to escape us. The knowledge-power nexus of the modern era is based on a hylomorphic model of relations. Simondon clearly explained how the hylomorphic model of the relation between form (abstract and active) and matter (concrete and passive) that governs our thinking about technology is applied at all levels of Western thought. When translated into social terms, for example, hylomorphism takes on a distinctly pyramidal appearance: active form (the idea of genius) actively determines passive, indeterminate matter (the workforce), between the one who commands and the one who obeys: "Form corresponds to what the man in charge has thought out for himself, which he must express in a positive way when he gives his orders: form is therefore of the *order of what*

3 Original French extract: "[…] malgré les apparences, une civilisation du rendement, en dépit des apparentes libertés civiques qu'elle laisse aux individus, est extrêmement contraignante pour eux et empêche leur développement, parce qu'elle asservit simultanément l'homme et la machine; elle réalise à travers la machine une intégration communautaire contraignante. Ce n'est pas contre la machine que l'homme, sous l'empire d'une préoccupation humaniste, doit se révolter; l'homme n'est asservi à la machine que quand la machine elle-même est déjà asservie par la communauté" (Simondon 1958, p. 355).

4 Original French extract: "Nous voulons que les robots aient toutes les qualités que les maîtres recherchent dans leurs esclaves, les patrons dans leurs employés, les commandants dans leurs soldats, et nous vou-lons qu'ils n'aient ni leurs faiblesses, ni leurs manques, ni surtout cette tendance irrépressible à l'insubordination, à l'indépendance, et à n'en faire qu'à leur tête qui caractérise les travailleurs humains" (Dumouchel and Damiano 2016, p. 10).

can be expressed; it is eminently active, because it is what is imposed on those who will manipulate matter; it is the very content of the order, what governs it"[5] (Simondon 2005, p. 51). The modern machine is the perfect slave, finally vindicating this binary logic of command.

For a long time, the mechanistic paradigm led us to believe that knowing something through the determined sequence of its causes and effects was a sufficient condition for controlling it and making it produce a desired effect. However, what contradicts this mechanistic conception in certain contemporary artificial intelligence applications is the fact that the effects produced cannot always be reduced to the a priori knowledge we have of them. The result of the algorithm cannot be reversed in the abstract form thought up by the "man in charge". There is more and more play in the mechanism. Although the *gap* between the mechanism and its performance was, in the staging of The Mechanical Turk, nothing more than a ruse of reason by the majority over the minority, a *trick* that someone held the secret to, the same cannot be said of algorithmic performance, for which engineers themselves are often placed at odds, and for which there is no person or place where the secret can be discovered. In other words, it is not enough to expose the machine, to strip it of its deceptive veil, in order for us (but who exactly?) to finally see its truth, because its mechanism is partly silent when it comes to accounting for the effects it generates (Seaver 2019; Burrel 2016). Faced with technical systems that are as distributed and multi-scalar as machine learning, a certain mastery can only be regained through an overhaul of public institutions and our technical culture. It is not our intention here to provide clues on how to tackle such a daunting challenge, although we have explored some of them elsewhere (Grosman and Reigeluth 2019; Reigeluth 2022). We merely present a preliminary question: if we fear losing control over AI, then what kind of pleasure do we derive from the feeling of being in control, and what other pleasure might take its place if the former turns out to be a source of frustration and fear – in other words, of sad emotions – rather than a power to act?

5 Original French extract: "La forme correspond à ce que l'homme qui commande a pensé en lui-même et qu'il doit exprimer de manière positive lorsqu'il donne ses ordres: la forme est donc *de l'ordre de l'exprimable*; elle est émi-nemment active, parce qu'elle est ce que l'on impose à ceux qui manipuleront la matière; elle est le contenu même de l'ordre, ce par quoi il gouverne" (Simondon 2005, p. 51).

9.3. An endgame

To try and answer this question, let us turn to what is mechanical in us and hypothesize that the pleasure we experience in controlling machines is perhaps, above all, a way of repairing our relative inability to control ourselves, an experience that is part of our psychic and collective life. The paleoanthropologist of technology André Leroi-Gourhan (1965) described the "programs" of everyday gestures as "mechanical" to distinguish them from purely unconscious automatisms and lucid behaviors that require sustained attention (pp. 28–29). While these behaviors are learned through the triple influence of imitation, trial and error and communication (Leroi-Gourhan 1965), their typical efficiency remains below the level of consciousness in the regular course of activity. The vast majority of our everyday behaviors plunge us into a "mental penumbra", which we only emerge from in the event of an obstacle or mishap. In other words, intelligence is never a unidirectional relationship between a command function and an execution function, but a ternary back-and-forth relationship between embedded unconscious automatisms and moments of lucidity and control. Intelligence is an in-between, which is located neither in the worker or the boss nor in the machine or the user. It is a living relationship that never flows in just one direction. This is why the promises of an AI found only in objects (however connected they may be) are the promises of a dead intelligence. Although this reticulation of algorithmic objects is part of the material and infrastructural conditions of our intelligence, it also assumes their integration into a social activity that gives them a purpose (Reigeluth 2018). If algorithmic performances are beyond the control of the engineer or programmer, and if they display levels of unpredictability and resistance, this is not so much an admission of weakness on the part of human reason, but rather an opportunity for a new practice in intelligence; a practice whose function would not be to re-establish unilateral control, but to bring a disparity of purposes into play. This was already pointed out by the Renaissance historian of technology Jonathan Sawday (2007, p. 111): "Play and fantasy [...] have perhaps been far greater elements in the evolution of different forms of technology than is suggested by the that popular (but wrong-headed) belief that 'necessity is the mother of invention'". We need to start focusing on how the behaviors of machine learning creep into our daily activities and participate in our own behaviors through the triple influence of imitation, trial and error, and communication. Desiring machine learning also means coming to terms with the idea that such a machine could disobey, pretend and circumvent the purpose we set for it. This possibility, which we come across in practice, could be a condition for exercising our intelligence if it were within reach of anyone, not just those with legitimate knowledge (and therefore control). The criticism here would then not so much be that "AI is naked!", but that anyone who participates in making AI effective beyond the person who thought it up or designed it (the user who "offers" their data, the worker who repairs datacenter breakdowns, the miner who extracts lithium and cobalt, etc.) can claim to dress it up as they see fit.

9.4. References

Bachimont, B. (2010). *Le sens de la technique : le numérique et le calcul*. Les Belles Lettres, Paris.

Burrell, J. (2016). How the machine "thinks": Understanding opacity in machine learning algorithms. *Big Data and Society*, 3, 1–12.

Canguilhem, G. (2009). *La connaissance de la vie*. Vrin, Paris.

Cardon, D. (2013). Dans l'esprit du PageRank. Une enquête sur l'algorithme de Google. *Réseaux*, 1(177), 63–95.

Cardon, D. and Casilli, A. (2015). *Qu'est-ce que le Digital Labor*. INA Editions, Bry-sur-Marne.

Collins, H. and Kusch, M. (1998). *The Shape of Actions: What Machines and Humans Can Do*. MIT Press, Cambridge.

Crawford, K. (2021). *Atlas of AI*. Yale University Press, New Haven.

Daston, L. (1994). Enlightenment calculations. *Critical Inquiry*, 21(1), 182–202.

Dumouchel, P. and Damiano, L. (2016). *Vivre avec les robots*. Le Seuil, Paris.

Eglash, R. (2007). Broken metaphor: The master–slave analogy in technical literature. *Technology and Culture*, 48(2), 360–369.

Grosman, J. and Reigeluth, T. (2019). Perspectives on algorithmic normativities: Engineers, objects, activities. *Big Data and Society*, 6, 1–12.

Kang, M. (2011). *Sublime Dreams of Living Machines*. Harvard University Press, Cambridge.

Kant, I. (2000). *Critique de la faculté de juger*. Flammarion, Paris.

Kant, I. (2008). *Réponse à la question : Qu'est-ce que les Lumières ?* Flammarion, Paris.

Lee, P. (2016). Learning from Tay's introduction. *Official Microsoft Blog*.

Leroi-Gourhan, A. (1965). *Le geste et la parole, tome II: la mémoire et les rythmes*. Albin Michel, Paris.

Poe, E.A. (1836). Maelzel's chess player. *Southern Literary Journal*, 2, 318–326 [Online]. Available at: http://www.eapoe.org/works/essays/maelzel.htm.

Reigeluth, T. (2018). La prédiction algorithmique comme activité sociale. *Réseaux*, 211, 35–67.

Reigeluth, T. (2022). Le rapport magique à l'Intelligence artificielle, ou comment vivre avec l'aliénation technique. *Quaderni*, 105, 35–52.

Reigeluth, T. and Castelle, M. (2021). What kind of learning is machine learning? In *The Cultural Life of Machine Learning*, Roberge, J., Castelle, M. (eds). Palgrave-Macmillan, Cham.

Sadowski, J. (2018). Potemkin AI. *Real Life Magazine*, 6 August.

Sandvig, C. (2016). When the algorithm itself is a racist: Diagnosing ethical harm in the basic components of software. *International Journal of Communication*, 10, 4972–4990.

Sawday, J. (2007). *Engines of the Imagination. Renaissance Culture and the Rise of the Machine*. Routledge, New York.

Schaffer, S. (1994). Babbage's intelligence: Calculating engines and the factory system. *Critical Inquiry*, 21(1), 203–227.

Schaffer, S. (1999). *The Sciences in Enlightened Europe*. University of Chicago Press, Chicago.

Schaffer, S. (2016). Babbage's dancer. *Imaginary Futures* [Online]. Available at: http://www.imaginaryfutures.net/2007/04/16/babbages-dancer-by-simon-schaffer/ [Accessed February 2016].

Seaver, N. (2018). Captivating algorithms: Recommender systems as traps. *Journal of Material Culture*, 24(4), 421–436.

Seaver, N. (2019). Knowing algorithms. In *digitalSTS: A Field Guide for Science and Technology Studies*, Vertesi, J. and Ribes, D. (eds). Princeton University Press, Princeton.

Simondon, G. (2012). *Du mode d'existence des objets techniques*. Aubier, Paris.

Simondon, G. (2013). *L'individuation à la lumière des notions de forme et d'information*. Éditions Jérôme Millon, Grenoble.

Winner, L. (1977). *Autonomous Technology. Technics-out-of-Control as a Theme in Political Thought*. MIT Press, Cambridge.

PART 4

Politics and Technology

Introduction to Part 4

Politics is no stranger to technology and is equipped with its own tools. Historically, the latter have been used to meet the needs of empires and monarchies, such as the tablets of the Sumerian empire, which are the first sources of writing and accounting available to us. In ancient times, democracy was a radical exception (though not to the extent of including women), and the republic was a somewhat less rare form of power-sharing and transparency. Along with Sun Tzu's *The Art of War* and Machiavelli's *The Prince*, the 11th and 16th centuries were true manuals for seizing and retaining power. In the Age of Enlightenment, following Locke, political technology shifted its attention to questions of decision-making legitimacy and the balance of power (Montesquieu, Hume). The main instigator of this moment was constitutional monarchy; the American and French revolutions, as well as Rousseau's thoughts, with his direct democracy and his legislative capacity, were an emerging but still minor movement. As the democratic foundation was gradually consolidated, the 19th century paved the way for a more rational approach to technical administration (Comte). The 20th century extended the scope of these ideas, and in the wake of Keynes' thinking, led to the idea of a welfare state, capable of ensuring the well-being of humanity through work and the fair distribution of economic benefits. The technical equipment of states has only grown over the decades, with a host of increasingly sophisticated and internationalized statistical and budgetary tools and indicators.

Introduction written by Valérie CHAROLLES.

All these tools were based on a concept of power that has been upset by the emergence of digital networks.

Indeed, the political thought of the Enlightenment largely revolved around the idea that technological development and democratic strengthening would go hand in hand: the fact that power belonged to citizens and emanated from them would enable us to move toward a less violent, fairer society; new technology would then provide us with the means for this shared ease in a virtuous circle of mutual understanding, as emblematically illustrated by Diderot and d'Alembert's *Encyclopédie*. Politics was thus at the heart of innovations, whether it was a question of bringing them to fruition (such as Turing and computer science in the context of the "Enigma" project) or enabling their implementation (railway and electricity infrastructures, etc.), which politics then mastered.

With the advent of digital networks and the growing power of information technology, the landscape has been radically transformed: the prevailing impression is that of the dispossession of political power by the major digital players and the powerlessness of politics in the face of technological change and the discourse that surrounds it. At the same time, however, states – including democratic ones – have made use of the unprecedented surveillance capabilities that these technologies provide. Both movements are a test for democracy.

The discreet surveillance exercised by states and certain economic players is currently jeopardizing democratic practice. The capacity for political decision-making is also being called into question, as it is now so tightly controlled by quantitative standards and performance indicators. Ultimately, the predominance of economic issues over social issues, as well as the imbalance between the powers of public and private players, is testing the very role of politics.

How can we overcome these challenges and integrate politics and technology? This can be done by realizing that technology, while reconfiguring our way of inhabiting the world, is also the result of human labor, and not an external issue beyond our control. This part of the book suggests ways of seizing our individual and collective capacity for action.

Pierre-Antoine Chardel opens this reflection with the idea of political micro-gestures. In the chapter entitled "Controlling Digital Technologies: Between Democratic Issues and Social Demand", he analyzes the inability to act due to dematerialization, as well as the discrepancies between predictions from technology developers and the unexpected, sometimes even transgressive, practices that these technologies ultimately give rise to. He encourages us to take advantage of these discrepancies to deconstruct techno-discourses and defends digital micropolitics, which is capable of both effective action and pluralism.

Bernard Reber then focuses on the practice of civic technologies. In "Responsibilities System: Ethics of Civic Technology", he starts from the concept of responsibility as defined by Jonas and examines the limited promises of remote participation. He draws on Dewey and Simondon to define an ethic for a highly technologized world and to reconfigure patterns and habits of thought. Positioning himself in the field of political and moral philosophy, he shows how the development of deliberative experiments, using digital technologies, cannot do without reflection on the forms of attention (concentration, duration).

Lastly, Valérie Charolles presents the concept of the "decidable" to position politics and exercise our power over technology. In "From the Infinite Universe to the Reflexive System: Uses of Technology, States of Emergency and Decidability", she starts by analyzing the relationship between technical developments and exceptional events in order to characterize a new form of the world and the subject: not the closed world of the ancient past where the collective comes first, nor the infinite universe of the classics whose mechanics are opposed to the freedom of the thinking subject, but a reflexive system in which facts echo and reflect around rules of the game. She proposes renewed frames of thought for this paradigm, which she also defines as the "end of naturality".

Andrew Feenberg's conclusion to the book is an invitation to the democratic appropriation of technology. Although Habermas turned to the re-enchantment of nature, Andrew Feenberg takes Marcuse as his starting point and considers the link between technology, the economic system and forms of life. This path he mentions involves investing in technology and understanding it better. Highlighting the importance of life and the role of capitalism in the social construction of technology, he shows how to move beyond a one-dimensional technological vision. It involves making an ontological leap away from the now obsolete physicalist paradigm and deploying a new scientific and technical model. The productions resulting from our interactions with technology, the way in which objects are transformed and the ways in which rationality emerges within societies are not only subjects for study but also areas for political action. In order to respond to environmental concerns, Andrew Feenberg urgently calls for these areas to be occupied.

10

Controlling Digital Technologies: Between Democratic Issues and Social Demand

10.1. Introduction

Technologies today are often made to serve capitalist ends. Interestingly, these technologies may have opened a new portal for individuals who are learning how to grasp them, thus resulting in possibilities that are not limited to the logic of consumption. However, to open up such perspectives, we need to be careful about which major trends we wish to promote in the evolution of our societies. While digital technologies can be a source of diversion, the socio-philosophical approach to them still faces a number of challenges. Overall, digital expertise is still very limited (Guichard 2019). The lack of terms and conditions in the digital environment is even having tremendous social consequences on the balance of social life. Public institutions tend to simplify their administrative procedures, and services are therefore often dematerialized; yet, these extended forms of services (i.e. the digital form of services) should be questioned from an ethical point of view. A quick example of this is that it may lead to new forms of vulnerabilities, such as social exclusion. How can we deal with this? What democratic impulses are likely to accompany the deployment of technology in our lives?

10.2. Dematerialization leads to an inability to act

When a number of public services are managed exclusively by virtual means, there is a significant risk that certain categories of the population will

---

Chapter written by Pierre-Antoine CHARDEL.

be secluded. It is, for example, difficult to impose immaterial payment methods on people who do not wish for them, thus implicitly assuming that no social life would be possible without digital means. This vision of society's smooth running, with its underlying ideology of progress, does not consider how closing down services can have negative effects on the life balance of elderly and vulnerable people. In this respect, public policies to dematerialize many everyday services should be accompanied by the analysis and the prevention of inequalities generated by the imposition of all-digital technology and by the proposal of a mediation space. There is still time to collectively question the place we want to give people who have no access to the Internet, who have difficulty handling it and who are likely to be unable to act in their daily lives, but also in social invisibility[1].

Dematerialization actually creates a paradoxical situation: instead of simplifying access to public services, it can jeopardize the effectiveness of equal access to these services for all users by further distancing it from them. In a context of growing precariousness, people are increasingly turning to social organizations for help. While the advantages of dematerializing certain public services should not be called into question overall, it remains essential to offer a wide range of information methods and to maintain physical reception points. It is therefore ethically imperative to reflect on and implement a refined, reasoned framework encompassing digital technologies and other means of information embodied by people, faces and word of mouth. This is a massive issue because exclusion and precariousness appear to be two of the consequences of public policies which are too unambiguous and irreverent of the challenges of social differences, when this same technical system is supposed to be imposed on everyone in the same way, without a process of skill-sharing needing to be undertaken.

10.3. Technologies and their social practices

Our modernity, which has long valued a certain number of principles, such as autonomy and communication, is facing intense contradictions. This is because the neutralization of the spoken word undeniably produces a

1 Here and in the following paragraphs we take up some of the aspects we discuss in a previous work entitled *Socio-philosophie des technologies numériques. Éthique, société, organisations* (Chardel 2022).

decrease in social life. We are reminded of this decline when, for example, we have to deal with an automated service (banking, insurance, administration, etc.) that forces us to interact with systems and pronounce keyword phrases or to respond to commands by tapping keys on our telephone (Besnier 2012). We are all too familiar with these interactive moments of absurdity with a machine. It is to avoid such a rising insignificance that organizations with a public service mission are setting up means of direct communication and personalized remote assistance. For example, the organization of certain social services, even in a teleworking situation, makes it possible to help vulnerable people in their daily lives by carrying out, sometimes on their behalf and with all due precautions, dematerialized procedures that they are unable to carry out on their own. While the absence of physical exchanges means that not all requests can be met, the social link is nevertheless preserved. Moreover, we know that in healthcare services, the living word of a person in the flesh is irreplaceable. This is also where the work of ethics comes in, by giving us the means to distinguish between a simple technical ideology and digital technologies themselves, the latter always being the object of heterogeneous social practices.

The history of technology shows that there is always a gap between predictions from designers and the practices of those who appropriate the tools offered, often in unexpected, even transgressive ways. In this respect, what the American philosopher of technology Andrew Feenberg recalls about Minitel at the time of its launch remains highly instructive: the official notion of online communication propagated by telecoms emphasized the democratization of information flows. Almost immediately, however, hackers introduced pink messaging services, and Minitel became something quite different from the vision Simon Nora and Alain Minc had of it at its launch: "The service went from cold to hot in the blink of an eye. Cybernetics undoubtedly has a reductive vision of communication, but I do not think it is relevant to Minitel or the Internet. These two media have been issues of singular appropriation by users" (Feenberg 2012). In other words, technologies are not independent of the social practices that are constantly being invented, which elude the often-binary concepts that emerge from certain narratives. In this respect, the resistance to 5G deployment is far from being the result of a *syndrome amish*, as the President of the French Republic put it in December 2020 (Chardel and Frérot 2022). Such a

Manichean view would even run against the effort of discernment we must generate if we honestly wish to engage in a constructive inquiry about the deployment of this new technological standard. This effort would involve critical action that would not just be *against* it but would refer to the ambition of acting collectively with responsibility toward future generations.

Continuing with the 5G case, it is clear that its promoters have no intention of subjecting its deployment to any kind of controversy. Questioning the ethics of the present and the future is not part of their perspective. They are not interested in the contribution of participatory technology assessments (PTAs), which have long been practiced in some northern European countries (Reber 2016). There is a feeling that we are reliving what happened back in the 1960s with certain land-use plans. Yet today there is a growing demand from civil society to participate in decisions affecting technical development. For example, at the launch of the French tracking app StopCovid in spring 2020, multiple alerts expressed by researchers and scientists from all fields, over many disciplines, helped to create a form of counter-power in the public space, alerting citizens and politicians to the ambiguities of this app in terms of privacy (Chardel et al. 2020). The app was then revised by integrating more legal guarantees. This is a sign that the practice of disputing digital technology is becoming an increasingly strong social demand and one that will undoubtedly intensify in the future. There is a kind of structural effect at work here, insofar as our technical practices in the digital era are also familiarizing us with the dynamics of contribution and participation, while confronting us with proven forms of *surveillance capitalism*. A horizontal culture has clearly taken hold since we became hyper-connected and networked.

10.4. Deconstructing techno-discourses for a better life with technology

Despite growing demands for "democratic technology" (Feenberg 2004), techno-discourses continue to grow: operators and access providers promote their networks and coverage, capturing consumers' attention as much as possible. Attention-control technologies are in full operation and almost systematically – and very effectively – focus on the benefits of leisure, comfort, culture and entertainment. Why should we be perplexed by devices that are supposed to bring us greater comfort and pleasure? There are numerous techno-discourses in this respect, enough to make our "heads spin". We are also told of the benefits of 5G virtual reality headsets for watching sports, taking virtual tours of museums, watching TV series on our smartphones and so on. Sadly, these arguments run counter to the demands for meaning that people have in the search for more alternative lifestyles. The momentum behind the *disnovation* program (Fourmentraux 2017) is, in this regard, indicative of growing demand based on the objective of stimulating a different relationship to digital technologies, one that is both more critical and transgressive, beyond the dominant technical ideology.

In a way, the more we evolve in technically complex environments, the more we have to focus on the complexity of the human being. It is ethically imperative to avoid idealizing the flow logics that are supposed to make living together better. The circulation of data flows means nothing in terms of a common goal, if it is not backed up by the capacity to interpret it: in the context of global ecological and health crises, data are no longer just agents of scientific rationality, but are also major political and socio-technical pillars (Carmes 2020). However, the largest possible number of people need to be able to interpret these *data*, without falling into the trap of generalized addiction to an empire of signals to the detriment of signs (Chardel 2020). Here, we are thinking of the moment when machines start interacting more and more with one another, without subjectively involving us, and instead start making decisions for us in order to make our lives easier: the case of the connected fridge already speaks for itself.

Taking this a step further, if we know that we can geolocate a person via the signal emitted by his or her cell phone, it is our bodies themselves that could be more easily recognized and identified with the advent of 5G, ushering in the era of *connected bodies* (Khatchatourov et al. 2022): in this situation, we are thinking of the face, which is likely to be digitized and to send signals (like a QR code, etc.) to remote facial recognition systems in public spaces. However, no one knows when or by whom the sum total of data collected without our knowledge in hyper-connected environments could be exploited. We can already see how China is using it for purposes that are anything but emancipatory. We realize that such innovations represent a frightening step toward generalized surveillance, regardless of the nature of the political regimes involved, whether they be "democratic" or openly "authoritarian". There is every reason to believe that we will become increasingly accustomed to being identified, even if the institutionalization of such an act is highly questionable from an ethical point of view, as it will no longer allow individuals to escape control, which we can still do today without a cell phone. With remote biometric recognition systems in the hyperconnected city, such a micro-gesture will simply become impossible.

10.5. Digital micropolitics

Faced with the diversity and complexity of the socio-philosophical issues raised, which essentially occur when 5G controversies are brought to light, could one ambition not be to limit the uses of the latter? Or to develop it in a more gradual manner? Could this be achieved by denouncing communication policies that are as deafening as they are alienating? Or, in particular, by avoiding its rapid spread into the public sphere? After all, what do we gain from trying to manage our lives ever more quickly? Not only are natural resources finite, but human and cognitive resources are not inexhaustible either. We cannot overlook the fact that digital technology is not just a tool: it is an environment in its own right, affecting the way we live, think and interact.

A gradual deployment – under democratic control – with rigorous public assessment should be the rule for all technical innovations. In this respect, a very interesting micropolitical phenomenon is emerging: the dynamics of controversy are emerging within public institutions and territories. Some local communities (cities and conurbations, in other words, fairly large institutions) are calling for the 5G deployment to be "paused", so as to enable controversy and in-depth debate (along the lines of the Citizens' Climate Convention) "to be organized within communities, at the city hall, community hall or regional hall" (Chardel and Frérot 2022). If such debates take place in earnest, institutions will be able to shift from a technical ideology to hospitality, welcoming controversy about the benefits or otherwise of certain technical deployments.

One of the key issues here is to move away from a fixed representation of progress decided once and for all, as if the history of our societies had already been written. It is also up to us to learn to live more *democratically* with digital technology. In order to avoid overly fantasized views of innovation, which would inevitably mean *better* views, from a democratic perspective, it would seem important to reflect on the conditions for an education that is more transdisciplinary and applied to technical complexity: a techno-ethical challenge is thus essential, and it is already being expressed in the search for economic models other than those proposed by GAFAM.

A number of initiatives are already taken in this direction, notably the Institut de recherche et d'innovation (IRI)[2] in France.

10.6. Promoting pluralism

We can see the extent to which the objective of combining ethical responsibility and technical innovation is now sociologically more important. Digital public policies should be designed to meet these requirements and thus embrace pluralism in our relationship with innovation. In particular, we need to enable transdisciplinary training, so that technologies are not viewed solely from the angle of pure instrumental rationality, but can instead address the emotional side of individuals, as in the case of artists creating sensitive digital devices[3]. In this respect, France is lagging behind northern European countries[4] in terms of education, given the disciplinary boundaries (between engineering sciences and humanities and social sciences) that persist on a massive scale and the difficulty of developing pedagogical perspectives that integrate artistic, scientific, technological and philosophical aspects, even if initiatives in this direction have been emerging in recent years. This margin for improvement is far from negligible, as our digital environments call for the development of new capacities for action from citizens themselves, who are still too often left out of decisions affecting the deployment of technology in their lives.

10.7. Conclusion

The development of popular digital universities, which already exist in some parts of France, would be a way of facilitating more enlightened, contributive appropriation of technical innovations. In this perspective, it is important to encourage the creation of places that encourage the meeting of opposites, as well as the creation of collective arrangements (Deleuze and Guattari 1980). Everything here remains to be constructed, bearing in mind

2 The French Institute of Research and Innovation.

3 Since 2011, the Marshmallow Laser Feast (MLF), a collective of immersive artists, has been offering the chance to step into the shoes of animal or organic creatures to establish new connections between humans and nature. See: https://usbeketrica.com/fr/article/quand-la-realite-virtuelle-nous-transforme-en-arbre-pour-nous-faire-aimer-la-nature.

4 See: https://www.weforum.org/agenda/2019/05/how-finland-is-fighting-fake-news-in-the-class-room/.

that one of the difficulties is to be able to join forces with heterogeneous social and economic environments, which are often kept at a distance from initiatives that could nonetheless bring them closer together. Many FabLabs are still relatively inaccessible to the ecosystems that surround them and therefore exclude a whole section of the population. However, beyond these limits, the momentum created by these dynamics of collaboration and this sharing of expertise is setting in motion a process of economic re-appropriation of spaces and territories. This is happening with the knowledge that a territory is not just a physical reality: it is always likely to open onto a world of shared skills, expertise and heterogeneous cultures. Behind these dynamics of reterritorialization, we have emphasized that forms of emancipation are multiplying and manifesting themselves in often highly innovative approaches, the characteristics of which express a desire to change certain paradigms by helping to free digital technologies from harmful economic models based on the commercialization of personal data.

10.8. References

Besnier, J.-M. (2012). *L'homme simplifié. Le syndrome de la touche étoile*. Fayard, Paris.

Carmes, M. (2020). *Les Cahiers du numérique*. Lavoisier, Cachan.

Chardel, P.-A. (2020). *L'empire du signal. De l'écrit aux écrans*. CNRS, Paris.

Chardel, P.-A. (2022). *Socio-philosophie des technologies numériques : éthique, société, organisations*. Presses des Mines, Paris.

Chardel, P.-A. and Frérot, O. (2022). Controverse sur la 5G. *Études. Revue de culture contemporaine*, 4289, 57–66.

Chardel, P.-A., Charolles, V., Delmas-Marty, M., Mhalla, A. (2020). StopCovid : il est contre-productif de proposer une solution techniciste à un problème qui ne l'est pas. *Le Figaro*.

Deleuze, G. and Guattari, F. (1980). *Mille plateaux : capitalisme et schizophrénie*. Éditions de Minuit, Paris.

Feenberg, A. (2004). *(Re)penser la technique. Vers une technologie démocratique*. La Découverte/MAUSS, Paris.

Feenberg, A. (2012). La pensée de la technique. Pour une approche humaniste. *Esprit*, 390, 57–58.

Fourmentraux, J.-P. (2017). Disnovation. *Techniques and Culture*, 67, 280–285.

Guichard, E. (2019). Plus qu'une ressource, le numérique est une forme violente d'exploitation. *Ouishare* [Online]. Available at: https://www.ouishare.net/article/le-numerique-va-vite-et-sans-delicatesse-plus-quune-ressource-il-constitue-aujourdhui-une-forme-violente-dexploitation-entretien-avec-eric-guichard.

Khatchatourov, A., Avenati, O., Chardel, P.-A., Queval, I. (2022). *Corps connectés. Figures, fragments, styles*. Presses des Mines, Paris.

Reber, B. (2016). *Precautionary Principle, Pluralism and Deliberation*. ISTE Ltd, London, and John Wiley & Sons, New York.

11

Responsibilities System: Ethics of Civic Technology

11.1. Introduction

This chapter presents an approach based on the work of the renowned phenomenologist and proponent of a biological philosophy Hans Jonas in favor of an appropriate ethics for a technological civilization. It then compares networked digital technologies (*civic tech*) with older technologies that are often thought to be rhetorical, literary and contemporary with the complexification of democracy. Although they date back to antiquity, they are still underdeveloped in today's innovative debates. This chapter also responds to some of the expectations we might have of a philosophy of technology, which are, at first sight, absent from philosophical debates on democracy and citizenship. It is not just a matter of framing technologies politically, but of also considering the part they play in democracy[1]. Similarly, thinking about technology can lead us to revisit forms of philosophical thought, as we present with Jonas and especially Dewey and Simondon.

Some typical cases of broad democratic participation, which were intended to be innovative and took place in France (2019–2022)[2] and Europe, can illustrate this reasoning.

Chapter written by Bernard REBER.

1 For the co-dependence of ethics and participation, at all stages of the research and innovation process, see Giannelos et al. (2022).

2 Available at: https://granddebat.fr/ et https://www.conventioncitoyennepourleclimat.fr/. For Europe, see: https://futureu.europa.eu/?locale=fr.

11.2. Improvisations on Jonasian responsibility

This chapter improvises with different leakage points on the terms of Jonas' (1991) deceptively simple book title, *The Imperative of Responsibility: In Search of an Ethics for the Technological Age*[3].

A first improvisation is the addition of a "responsibilities *system*". The addition of the system, and not just responsibility in the singular, gives way to *responsibilities*. This means that they interact within a system, and they can be more or less *responsive* to each other. It could even be said that, for major issues such as the ecological transition or technologies with widely dispersed impacts over long periods of time, these responsibilities are both *common and different* depending on the actors involved (Reber 2016a).

These additions to Jonas' work contribute to the transition that this creative philosopher never made: from ethical reflection, or even meta-ethical reflection or moral anthropology for him, to questions of philosophy and public policy. Indeed, he considered this transition too difficult (Reber 2016b). Paradoxically, with his "creeping apocalypse of technologies", Jonas does not give us the means to control them and be responsible for them.

However, even on the meta-ethical level, with responsibilities in the plural, we go further than Jonas with a second improvisation. He does not consider moral pluralism, which is nevertheless guaranteed in liberal democracies (Rawls 2003; Reber 2016b), nor the layering of common and differentiated responsibilities, which must be able to be organized into a system. This pluralism leads to necessary deliberative democracy (Bächtiger et al. 2018; Reber 2023) and not just mini-publics, distracting us from the bigger picture; deliberative democracy is first and foremost a *deliberative system*.

We can mention several paths and pitfalls in France to strengthen our remarks, from the Yellow Vests (*Gilets jaunes*) crisis, which, through its

3 In French, the term *"Principe"* (principle) is used instead of "imperative", like in German.

actions, benefited social networks of major capitalist companies such as Facebook, with limits and ambiguities, to France's Great National Debate (2019), the Citizens Convention for Climate (2019–2022) and the reinterpretations of the French Climate and Resilience Law discussed in Parliament. This did not happen in one single stage but was followed by all sorts of contingencies, such as campaign promises by candidates to dismantle wind turbines or the aim to reduce greenhouse gases (GHGs) in the European Union by 50%, well above the 40% imperatives of the French Citizens Convention for Climate (Courant and Reber forthcoming), not to mention the various legal disputes with the government in France for having revised its climate ambitions downwards. All these actions by very diverse players, with equally diverse capacities for action and responsibilities, are part of a system of systems whose coherence is an enigma. This difficulty has already been faced by the historical sciences of the past, but it is hard to say what the trajectory of the future will be. We could have added an unexpected event to this convention, which came along and allowed the audacity of State spending, the "whatever the cost" approach, with Sars-Cov2. This had an impact on both the course of the convention, which was shifted to a remote method, and on the absence of arbitration on economic choices to finance the ecological transition.

Rather than the deliberative system alone, which is currently only scattered across several political science research programs (Bächtiger et al. 2018), we think that a system of responsibilities is more appropriate. This is not just true for the climate but for all themes, including those discussed during France's Great National Debate (taxation, state reform, citizenship and democracy, ecological transition) or, even more broadly, during the Conference on the Future of Europe, which tackled all European policy issues.

Jonas' sub-title, *In Search of an Ethics for the Technological Age*, has also been modified for a third improvisation. Indeed, we believe that technological civilization is a pleonasm. There is no civilization without technology. In fact, archaeologists use technical variations that can be spotted in the traces they uncover to classify them into particular periods.

Finally, a fourth improvisation: technologies play a part in education and civic life. For example, "citizen" comes from *civis*, the one who has the right to live in the city (*civitas*): the city with politics, which comes from *polis*. However, cities and towns are concrete technical networks occupying territories. Democracy has its share of technology.

It is also, as we shall see, a sophisticated technology for manipulating abstract notions (Dewey 2003; Schudson 2012; Reber 2023). These technologies are also civic technologies, far more flexible than connected cloud resources alone. With these civic technologies, we could even hijack the project of Latour and Weibel (2005), curators of an associated exhibition presenting all forms of parliament called *Making Things Public*. This exhibition sought to rediscover the materiality of the thing, the *res*, and the public, *publica*. It echoes research into the philosophy of technology during the thingly turn (Verbeek 2005)[4]. Some technologies are concrete, others are more abstract. Concreteness is not always the most important thing when it comes to navigating the order of discourse required for democracy.

11.3. Civic technologies

Democracy as a system is often a better guarantee than the habits of the individuals – brains and bodies – involved, particularly when it comes to openness to pluralism and contradictory criticism (Reber 2023), or simply to the very disparate information to be arranged. There can be no politics, especially for complex societies with a social division of labor, without trans-contextualization, whether between territorial scales, types of skills or particular contexts, such as our own. Yet in addition to institutional systems, since the invention of writing, literary technologies and communicative capacities, rhetoric composing the order of discourse, have had to accompany this complexification. Sophisticated communicative capacities, such as narration, interpretation, argumentation and even reconstruction (Ferry 1991a, 1991b, 2004), are involved in describing the world, describing what is desirable and, above all, in assigning responsibility. These capacities are associated with the ability to criticize. Democracy is the only system that accepts self-criticism. As a result, it is constantly changing. It is even a machine for deception, in accordance with its ideals. These communicative capacities are precisely what enable us to move from one context to another. A narrative produced in one context and reread in another requires interpretation. This sometimes involves a conflict of interpretations.

4 See Chapter 2 of this book by Xavier Guchet.

Likewise, interpreting a situation in terms of justice may require an even more universal approach and the need for argumentation. What the lawyer does conscientiously and with great technicality, by qualifying narratives with legal categories is much more unclear in political discussions. Nevertheless, such clarifications would be welcome.

In the title of this chapter, "Responsibilities System: Ethics of Civic Technology", we have put "ethics" in the plural. Indeed, moral pluralism requires values, but also theories (Reber 2016b). We therefore also need ethical participation rules (Giannelos et al. 2022) in institutional design choices, and therefore also assessment criteria, but, above all, we need an answer to this missing part, the most forgotten part, one that arises from the very first discussions around a table or in a hemicycle or even in a Yellow Vest traffic circle: how do we talk to one another? Or even more difficult: how do we talk about these abstract things? This was the major obstacle faced by the makers of a film about the Citizens Convention for Climate: how do you reveal and discuss what cannot be filmed? They had to fall back on the usual framework of participant portraits (Courant and Reber forthcoming).

With political debate, we are therefore dealing with the abstract, the order of discourse, but also with the rehabilitation of the spoken word. We are rediscovering techniques as old as Aristotelian naturalism, corroborated by very contemporary works: humans are rhetorical (Danblon 2013; Reber 2023). We are rhetorical because we are political, and rhetoric has enabled us to keep up with the evolution of other techniques in order to live together[5]. With rhetoric, we hold both institutions and public discourse. We think about action, expressing emotions, telling the story, suspending judgments, deciding and judging, to quote the philosopher of argumentation Emmanuelle Danblon in her book *L'homme rhétorique. Culture, raison, action* (2013).

Other technologies have their part to play, expected to be technical and much talked about, technologies boldly described as "civic". However, what kind of civility do they represent? Do they, above all, protect deliberation, the first type of distancing that Pierre Caye refers to in this book? We cannot be sure. We are dealing with an overflow of overly open access, often accumulating very short attention spans; in the case of France's Great

5 Those of "type 2", which Pierre Caye discusses in this book. We could just as easily have quoted Murray Bookchin and his concept of institutions (Chardel and Reber 2014).

National Debate (2019), the documents deposited on the platform by almost 2 million Internet users have been entrusted to the French National Archives, scattered all over the country, complicating the work of researchers or citizens who would have liked to consult the participants' views. What exactly did they participate in? A great deal of "data" were collected, but they were not given – or given back – when expected to in order to develop the regional citizens' conferences of France's Great National Debate (2019), which were the last phase of the debate, with systems made up of 100 or so citizens chosen by sortition in all the regions of France, as well as the overseas territories and departments.

Indeed, it would take a lot of processing time to make sense of these views, even with the use of so-called artificial intelligence (AI), especially with expectations structured on the basis of the communicative capacities developed above. Argumentation on a civic platform is not just about likes or the repetition of claims in the form of brief comments.

Furthermore, if we end up on this platform with the juxtaposition of contradictory political injunctions – we find everything and its opposite – we are not much further ahead. Even the President of the Republic, whom journalists have identified as the king of "en même temps" ("at the same time")[6], was unable to make the most of it. With the coexistence of expressions without criticism, without judgment, without renunciation and without the art to achieve it, we only have ingredients. Cooking is still necessary, or to put it less metaphorically: we must use reason.

Information and sometimes communication technologies (ICTs) certainly enable unprecedented connections, a form of remote inclusion, but they also tend to *disrupt*, in the less complimentary etymological sense of "disruptive" and sources of upheaval.

11.4. The limited promise of remote participation

Now is time for participation, whether face-to-face or via the civic tech platforms we have described above. These are literally civic technologies. For those with modest means, they are supposed to gather the views of

6 It is surprising that this expression is attributed to Emmanuel Macron, since politics in a democracy is always about "at the same time", in the face of pluralism and the need to build compromise.

participants, which can sometimes be very numerous. More rarely, these platforms enable exchanges and even more or less structured conversations between citizens. We are often dealing with chatter rather than political conversation (Dewey 2003; Schudson 2012, 2012b; Reber 2023). Communicative exchanges, with the diversity of communicative capacities mentioned above, ranging from narration to argumentation, very quickly come up against certain limits. There is a disparity of attention. You may only spend a short time on the platform, or you may be on it for a long time. You obviously cannot read or respond to everything. Algorithms can certainly structure parts of conversations or, more often, chats, but some of these will break away from what a real conversation in a shared space would allow, in addition to having a deliberative aim. Broad participation and extensive communication are therefore, from a certain point of view, antagonistic.

Voting is one way of getting around this difficulty. It is also known as "giving a voice". However, the voice, though a singular marker, is homogenized by voting. By giving your voice, you lose your timbre. Furthermore, with voting, all voices are equal. The voice of the voter expressing his vote stops at the communication threshold. ICTs share a form of confinement with the polling booth, not to mention isolation. The Internet user, like the responder to an electronic survey, is alone in front of their screen[7]. They escape questioning and challenges, and even the pressures of real, participatory discussion; similarly, they are not entitled to the richness of real interactions, both between the subjects exchanging views and for those listening to them.

It is no secret that people insult one another on networks hastily referred to as "social" networks. With all their attention-grabbing excesses, the widespread lack of civility and the absence of faces (Chardel and Reber 2001), they should actually be called "asocial" networks. Real interaction questions are expressed on faces. The rise of emoticons is no stranger to this need to associate messages and emotions or sentiments, something that rhetoric has long known (Danblon 2013; Reber 2023). That being said, our faces are better than emoticons.

7 It is well-known to surveyors that answers given in front of a screen or face-to-face can vary, especially for opinions that are more difficult to express, such as an intention to vote for an extremist party.

However, it is questionable whether these debates, being remote or face-to-face, live up to expectations. To find out, we need to have an idea of "how to debate" and "how to speak". In response to the "why?" question of participation, understood as the objectives pursued, we often find concerns about the legitimacy of the decisions made, their elaboration, implementation and effects. Nevertheless, the organizers of these debates, which are expected to be participatory or even deliberative, are rather brief on the communicative capacities to be supported. They have scarcely explored the frameworks of conversation and even less that of consideration (Reber 2023). They do little to clarify the various responsibilities involved or the exchanges about them.

Even when it comes to "how to participate", they have not considered the difficulties of their own judgment against the judgments of others, conjugated in the plural. The real problems of judgment difficulties, well presented by Rawls (Rawls 2003; Reber, 2013a, 2016b), are very rarely taken into account, even though this is one of the points that establishes the recognition of pluralism and the coexistence of critics.

11.5. Contributions of the philosophy of technology

What can we therefore expect from philosophies of technology? The question is too vast, both in terms of technology and philosophy. These fields are too heavy and diversified. The present book offers a guide for reading between these worlds. Additionally, philosophy, in certain respects, is a technology. It mobilizes technologies and has methods. Certainly, its tools are abstract. However, some concepts owe their etymology to tools. This is true for the *balance* of deliberation, or, in moral philosophy, the reflective equilibrium that was so dear to John Rawls[8].

Several chapters in this book have handed the fate of technology to conventions, understood more broadly like the humanities and social sciences think of them, or even leading metaphysicians, such as David Lewis (1969), who wrote his dissertation on the subject of convention. However, conventions can be underdeveloped or disputed when technologies are emerging or controversial. We are focusing on moral and political

8 For an approach to the debts of thought to technology, see Brossaud and Reber (2007a, 2007b, 2010).

philosophy, which is already a huge undertaking. Entire areas of innovations, which could be qualified as civic, have emerged. In traditions such as participatory technology assessment (PTA; Reber 2011), which have shaken up the field of institutional innovation, then the field of innovation and responsible research (IRR; Gianni et al. 2018)[9] and all forms of open science and participation (Giannelos et al. 2022), we eventually go through an assessment, a reflexive trial, when it is not a court of reason as in Kant's case or a courtroom in the case of damages to be repaired. Once again, with the courtroom, we have a technology in all its materiality, a debt of thought to technology (Brossaud and Reber 2007a, 2007b, 2010). However, as with rhetorical technologies, the laws of the courtroom are made elsewhere, in parliament. In the case of technologies, elected representatives have sometimes delegated these technological choices and assessments, where ethics are disputed, to citizens' conferences (Reber 2011). There, the "ought-to-be" of these objects is judged, as well as their processes, uses, and even any form of ideological or economic hypostasis where technology is recklessly approached as an autonomous subject, right up to its participation in the architecture of the world, whether desired or fought for.

Our book *Precautionary Principle, Pluralism, Deliberation: Science and Ethics* (Reber 2016b) makes it possible to think about these problems in ethical, scientific and political forms, with the challenging problems that their links presents. However, we could bear in mind the technologies themselves for thinking and not solely rely on the sub-domains of moral, political and scientific philosophies. In fact, some philosophers have traditionally taken technology seriously in various ways. This is particularly true of Jonas, but also of Dewey and, in an even more original way, Simondon (Reber 2008). Jonas' questions, and his call for an ethics of the future for a technological civilization (1981) in the face of a "creeping apocalypse" threat, could be answered by the political philosophy advocated by Dewey (2003). For Dewey, it is the investigation of the unforeseeable consequences of new technologies that, in a way, summons the public and calls for a State's constitution. Simondon's (1964) interest in the realization of techno-objects restores their place and nobility. His analyses and prognosis of the relationship between humans, the environment and

9 See the sets of books in "Responsible Research and Innovation" and "Innovation and Responsibility" (ISTE Ltd and Wiley): https://www.istegroup.com/en/series/innovation-et-recherche-responsables/ and https://www.istegroup.com/en/series/francais-innovation-et-responsabilite/.

technology are more optimistic than Jonas', since he believes that the phase shifts between human rhythms and machine rhythms, which generate alienation, will be rebalanced by cybernetics.

Dewey and Simondon do not simply propose ethics for a highly technologized world. Some of their more radical and original work (Reber 2008) takes the opposite path, such as when Simondon (1989) seeks an "ethics that would emerge from technology" (p. 177), or when Dewey presents it as "instrumentalism", for lack of a better term, as he acknowledged. The former challenges the old Aristotelian hylomorphic pattern, uniting form and matter, and the latter rejects the distinction between intrinsic and instrumental objectives, or even the separation between natural and moral sciences. Both also challenge an individualistic, atomistic conception of the act in the assessment and apprehension of moral issues. At the heart of their philosophy is a critique of ontogeny, which seriously considers the question of temporality, and thus the becoming of technology. Dewey and Simondon are not simply original philosophers of technology. It is this interest in technology that enables them to reconfigure (Dewey) and criticize (Simondon) patterns and habits of thought.

11.6. Conclusion

Civic technologies are just as important as ethical and political technology assessment, or simply ethics for a technological civilization, as Hans Jonas proposed. All civilizations are technological. Although the power of modern technologies is different and forces us to think more carefully about ethics, some of these technologies, which are digitized and connected, are supposed to reinforce the inclusion of democratic debate or participatory assessment. However, it is quickly forgotten that citizenship and democracy are technologies, and that they have had to invent forms of communication that rhetoric has given to us. We have therefore compared these two types of technology and shown how far civic tech still has to go to reach the requirements of the former.

Conversely, the development of deliberative experimentation using digital technologies has the advantage of requiring an explanation about what we want to do and the equipment we intend to develop. Choices have to be made, implemented, as it were, within the technical possibilities offered to participants. Of course, not all the anticipated uses will be fully determined, but the possibilities will be more or less extensive. A patient analysis of actual usage would still need to be carried out for recent platform experiments, backed up by large-scale debates. It will also have to be comparative, revealing the differences between face-to-face and remote exchanges. In both cases, the issues of time and attention will be essential. Technology may reduce distances, but it does not reduce the attention span required for rich, attentive communication. When attention is given – and not just one's voice – it is an element of consideration (Reber 2014, 2023). Real presence is not always real. *Transubstantiation*, to take up a theological dispute on real presence in the Eucharist, suggests this kind of remote, even digitized, presence. This is not to mention the entire phenomenological tradition, which has been able to grasp the weight of the present and apparition, and long before that, mystics of all schools: from philosophic spiritual exercises (Pavie 2020) to Taoists, all take temporality and concentration seriously in order to be placed in the presence of the present, as an exercise to be repeated over and over again.

We soon realize that presence and attention are variable, diverted by the flow of information. Therefore, there is a lot of absence. One might even lament a particular form of abstentionism. Containment experiments have enabled us all to grasp these differentiated forms of presence on screen or "face-to-face"; the term is not chosen at random.

The philosophy of technology can bring the same attention to technologies. It could emphasize an oversight in the famous example used to think about moral responsibility and its different concepts: the case of a sinking ship. This case is often presented in terms of the captain's responsibility. In the example, the captain has authority, skills and a role, exercised in a more or less virtuous manner, and is legally responsible for the shipwreck because they have been drinking. These five aspects of moral responsibility can be attributed to an individual. However, what all moral philosophers have forgotten in using this example is the boat. What is its part in the shipwreck? Is technical reliability not a candidate for concepts of responsibility? Indeed, for technology, the problem is not responsibility but reliability.

11.7. References

Bächtiger, A., Dryzek, J.S., Mansbridge, J., Warren, M. (eds) (2018). *The Oxford Handbook of Deliberative Democracy*. Oxford University Press, Oxford.

Brossaud, C. and Reber, B. (2007a). Conclusion. In *Humanités numériques 1. Nouvelles technologies cognitives et épistmologie*, Brossaud, C. and Reber, B. (eds). Hermes-Lavoisier, Paris.

Brossaud, C. and Reber, B. (2007b). Conclusion. In *Humanités numériques 2. Socio-informatique et démocratie cognitive*, Brossaud, C. and Reber, B. (eds). Hermes-Lavoisier, Paris.

Brossaud, C. and Reber, B. (eds) (2010). *Digital Cognitive Technologies: Epistemology and Knowledge Economy*. ISTE Ltd, London, and John Wiley & Sons, New York.

Chardel, P.-A. and Reber, B. (2011). Les risques de la communication assistée par ordinateur. *Communications*, 88, 149–157.

Chardel, P.-A. and Reber, B. (eds) (2014). *Écologies sociales. Le souci du commun*. Parangon, Lyon.

Courant, D. and Reber, B. (eds) (2024). *Convention citoyenne pour le climat. Démocratie délibérative et transition écologique juste*. ISTE Editions, London.

Danblon, E. (2013). *L'Homme rhétorique. Culture, raison, action*. Cerf, Paris.

Dewey, J. (2003 [1927]). *Le public et ses problèmes*. Publications de l'Université de Pau, Farrago, and Éditions Léo Scheer, Pau.

Ferry, J.-M. (1991a). *Les puissances de l'expérience. 1. Le sujet et le verbe*. Cerf, Paris.

Ferry, J.-M. (1991b). *Les puissances de l'expérience. 2. Les ordres de la reconnaissance*. Cerf, Paris.

Ferry, J.-M. (2004). *Les grammaires de l'intelligence*. Cerf, Paris.

Giannelos, K., Reber, B., Doorn, N. (2022). *Responsive Ethics and Participation. Science and Democracy*. ISTE Ltd, London, and John Wiley & Sons, New York.

Gianni, R., Pearson, J., Reber, B. (eds) (2018). *Responsible Research and Innovation: From Concepts to Practices*. Routledge, Oxford.

Jonas, H. (1991 [1979]). *Le principe responsabilité. Une éthique pour la civilisation technologique*. Cerf, Paris.

Latour, B. and Weibel, P. (2005). *Making Things Public. Atmospheres of Democracy*. MIT Press, Cambridge.

Lewis, D. (1969). *Convention*. Harvard University Press, Cambridge.

Pavie, X. (2020). *Critical Philosophy of Innovation and the Innovator*. ISTE Ltd, London, and John Wiley & Sons, New York.

Rawls, J. (2003 [2001]). *Justice as Fairness. A Restatement*. Belknap Press, Cambridge.

Reber, B. (2008). Quand la nouveauté technique oblige à penser autrement. Lecture de Hans Jonas, John Dewey et Gilbert Simondon. In *Phénoménologie et technique(s)*, Schmit, P.-E., Chardel, P.-A. (eds). Cercle Herméneutique/Vrin, Paris.

Reber, B. (2011). *La démocratie génétiquement modifiée. Sociologies éthiques de l'évaluation des technologies controversées*. Presses de l'Université de Laval, Quebec.

Reber, B. (2013a). Governance: Precautionary principle and pluralism. *International Social Science Journal*, 64(211/212), 75–87.

Reber, B. (2013b). Evaluation and promise of "e-democracy" around some consensus conferences. *Humanistyka i Przyrodoznawstwo [Humanities and Natural Sciences]*, 19, 153–164.

Reber, B. (2016a). Sens des responsabilités dans la gouvernance climatique. *Revue de Métaphysique et de Morale*, 89, 104–117.

Reber, B. (2016b). *Precautionary Principle, Pluralism, Deliberation: Science and Ethics*. ISTE Ltd, London, and John Wiley & Sons, New York.

Reber, B. (2023). *Responsible Deliberation, between Conversation and Consideration*. ISTE Ltd, London, and John Wiley & Sons, New York.

Schudson, M. (2012 [1997]). Why conversation is not the soul of democracy. *Critical Studies in Mass Communication*, 14, 297–309.

Simondon, G. (1964). *L'individu et sa genèse physico-biologique*. PUF, Paris.

Simondon, G. (1989 [1958]). *Du mode d'existence des objets techniques*. Aubier, Paris.

Verbeek, P.P. (2005). *What Things Do: Philosophical Reflections on Technology, Agency, and Design*. Crease, R.P. (trans.). Penn State University Press, Penn State University Park.

12

From the Infinite Universe to the Reflexive System: Uses of Technology, States of Emergency and Decidability

12.1. Introduction

Much has been written about the use of technologies during recent states of emergency[1]. Indeed, September 11 marked a break with a long-term trend where respect for the private sphere had become the norm, at least in theory, within the republican (Kant's *Towards Perpetual Peace*) and democratic paradigms (Declaration of Human Rights): unlike the panopticon described by Foucault in *Discipline and Punish*, surveillance technologies were only supposed to be used in exceptional circumstances. Since 2001, the situation created by the terrorist attacks has been used to legitimize the mass use of new surveillance technologies, which has been made possible by the digital revolution. The transmission of information and communications in packets of bits makes it possible to process and index data with an unprecedented scale and discretion.

The states of emergency linked to the Covid-19 pandemic, which began in 2018 and continued through 2019 and 2020, are even more overwhelming in terms of restrictions on freedoms and, in this regard, appear as a precedent. Whether or not this is just a phase, however, is still open to debate. The related technical facts are also powerful, but in a more bivalent

Chapter written by Valérie CHAROLLES.

1 For an exhaustive bibliography, please refer to Charlène Biondi's PhD thesis (2022).

way. The use of surveillance applications as indiscriminate tracing tools has been widely criticized[2] and has generally failed to gain acceptance; on the other hand, vaccination technologies have been deployed very fast and screening technologies widely used when targeted (tracing contact cases of sick people), making it possible, thanks to massive campaigns, to put an end to the state of exception in almost all countries after two or three years[3].

This chapter tackles this question by looking at continuities and disruptions in terms of technologies, political practices and philosophies, which can provide a framework for analysis. It achieves this around a classic definition of technology ("methods for using scientific discoveries for practical purposes", *Cambridge Dictionary*); it is not therefore limited to digital technologies: there are political technologies, essentially illustrated in the ways in which the law is passed, but also accounting technologies and so on. Technologies are not only visible tools in the form of an object.

What are the links between the deployment of technologies and that of exceptional events? What is the new form of the world at stake today, a world that could be described as post-natural? Does the Enlightenment frame of reference provide a suitable basis for addressing this world? What other foundations could legitimize and clarify the place of politics? These are the four questions we seek to clarify using the framework proposed in *Philosophie de l'écran* (Charolles 2013).

12.2. Deployment of technology and exceptional events

Historically, technical tools and practices have been deployed to tame nature, limit exceptional events or, more often, reduce their effects. A typical example is the introduction of the 3-year crop rotation system in the Middle Ages, a cultivation technique that improved yields by ensuring long-term soil regeneration, limiting the risks associated with climatic hazards. René Descartes' "make ourselves the masters and possessors of nature" (1637) offers a "clear and distinct" formulation of this objective, at a time when discoveries made through the invention of the lens were leading to a radical change in our conception of the world: we moved from the closed world of

2 See Chapter 10 by Pierre-Antoine Chardel in this book, as well as Chardel (2014).

3 At the time of publication.

the ancients to the infinite universe of classical science (Koyré 1957), an infinite universe that would become Newton's universe and serve as a foundation for Kant and the Enlightenment.

Nevertheless, exceptional events and the use of new technologies often go hand in hand, such as the atomic bombing of Hiroshima and Nagasaki that ended the Second World War in Asia. While the atomic weapon did not make war exceptional, it did respond to it in an exceptional way, illustrating the variety of uses that can be made of a scientific discovery. Unlike Turing's development of computers during Operation Enigma and the decoding of German transmissions, the discovery of the nuclear reaction was not linked to military ambitions.

In this long history, the deployment of statistical techniques marks an important point. Following in the footsteps of Blaise Pascal, frequency calculations were introduced in the early 19th century by Legendre, then Laplace and Gauss. The so-called "normal" law of Laplace and Gauss very quickly gave a canonical form to the way in which facts were supposed to be ordered (the bell curve): many average facts, very few extreme events, so few that their occurrence could be neglected in statistical calculations. This is a foray outside the Newtonian universe and into a world where absolute determinism is no longer the sole norm. It gives way to frequency laws and a new standard according to which facts occur in a "normal" way, allowing extreme observations to be scientifically set aside.

However, in this age of screens, Gaussian statistics are no longer a suitable depiction of how the world works (Charolles 2013). Computing power and the networking of information on computer screens break an essential condition of the Gaussian standard: the independence of observations. Search engines on the Internet or social networks are built on the opposite model, the linking of observation (the more you are searched for, the more you appear on the Web), as are financial markets. Computers and the networking of information have totally transformed their topography, replacing the brokers' trading floor with panels of screens linked to high-powered computers. More deeply, they have changed the way brokers work, with indexed and derivative instruments setting up large-scale echo and mirroring mechanisms. These practices have been legitimized by a new theorization of market efficiency according to which the best strategy is to follow the market. The results are feedback loops and fat-tailed distribution which, depending on the moment, lead to peaks or cracks, in short, to a multiplication of extreme events (Mandelbrot and Hudson 2005).

At the same time, the use of these tools by capitalism (which we distinguish from the market economy and economic liberalism (Charolles 2006, 2021)) and the theoretical foundations of the neoclassical school have based the economic standard on a logic of maximization, maximization of interest or utility by economic agents. This logic is very unlike the classical model, which is based on interest, but does not maximize or optimize it[4]. Thus, the highest growth is considered the epitome of economic success, even though classical economic thinking predicts a time horizon of equilibrium. Long history shows cycles (growth, plateau, recession, with a return to a state that is generally higher than that at the start of the cycle). It would therefore be quite exceptional for us to experience growth alone.

These discrepancies between average facts, expected facts and actual facts are all the more apparent as quantification has become the preferred way of qualifying facts, even when they involve humans. The deployment of statistics, computing power and the position taken by the economic field in social and political life have all contributed to this. We thus find ourselves faced with a new unifying problem, which is no longer that of the relationship between words and things put forward by Foucault[5] and Quine, but that of the relationship between facts and figures (Charolles 2008, 2022).

Hazards have always existed, but our technical tools have been deployed on such a scale in terms of tracking capacity and global coverage that we are now able to recognize exceptional events much more accurately than ever before (just think of the way in which the Covid-19 epidemic is monitored compared with the so-called Asian or Hong Kong influenza epidemics of the 1950s and 1960s).

This does not mean that technologies are responsible for the increase in crises since the 1970s[6] but that certain uses of technologies based on certain theoretical assumptions (normal distribution law, maximization of interest in economics) have led to extreme situations. Depending on the background, they can be seen as such (financial *krachs*) or not (stock market peaks,

4 The disruption came with Walras and Pareto.

5 The original title of *The Order of Things* was "Words and Objects" (*Les mots et les choses*).

6 See Dupuy (2002 and 2012) on the notion of "Enlightened Doomsaying".

uninterrupted economic growth). Indeed, while our equipment enables us to better measure exceptional events, its statistical foundations have been built around a narrow vision of normality. This vision maintains a sense of permanent crisis and can also prevent us from seeing other facts from the crisis angle. Our networked world accelerates these processes and their effects.

12.3. From the infinite universe to the reflexive system or the end of naturality

Is a new form of the world being implemented through the deployment of the cinema screen, the computer and networks; in other words, through our capacity to see images in motion, to have extremely powerful calculators at our disposal and to network them – an ability that barely dates back 120 years and was the stuff of dreams for both Leonardo da Vinci and René Descartes? Our answer is yes.

Alexandre Koyré (1957) showed how the invention of the astronomical lens in the 16th century enabled us to move from the closed world of Antiquity, with the flat Earth at its center, to the infinite universe of classical science. The discovery of the banality of the Earth in this mechanical universe has as its counterpart the affirmation of humanity as a radically different entity, the bearer of free will (Descartes) or of a moral law (Kant), and no longer primarily a political animal (Aristotle). The *screen*, to sum up our technical state, once again makes us change our frame of reference. We characterize it as a "reflexive system" and relate it, in the manner of Koyré, to a new dimension taken on by technologies (Charolles 2013). Indeed, it is not just the economic and entertainment spheres that the screen has changed. It has also changed the way in which we represent the world and, more deeply, the way in which it is constructed.

In terms of physics, new telescopic observation capabilities show us a universe that is far more complex than was previously assumed: it is no longer infinite and unchanging, but is now conceived as finite and endowed with a history[7]. At the same time, human activities are organized around increasingly dense networks that operate in an echoing, mirroring mode, such as world stock exchanges and social networks. On a global scale,

7 Based on our current knowledge, this history begins with the Big Bang and is situated in a finite space, like edges that can be circled around without limit.

civilization and "natural" reality are intertwined: human beings have reached such a degree of skill that they are no longer simply subject to the laws of evolution, but are, in a sense, its actors, having put themselves in a position to direct their conditions of reproduction and dominate the ecosystem. This ecosystem is organized in the form of a "reflexive system", in which the various strata of reality constantly interact.

It is therefore not just a form of representation of reality that is at stake, but deeper characteristics. While they may be in line with the diagnoses presented by Husserl's *Krisis* (1935) or the early Habermas (1968), these deeper characteristics go beyond that and have a metaphysical scope. Conceptions of space, time and relationship are involved.

In a century, all aspects of life have been taken over by screens. Our reference space has been totally modified at this time when relativity and quantum theories have shown that there is no univocal way of representing space and that indeterminacy exists at a certain level of observation.

With the invention of cinema, followed by television, computers and networks, the far-off world has become part of our living space. The first major mass transport ceremony was the broadcast of Neil Armstrong's first steps on the Moon in 1969. Since then, the spread of smartphones – miniaturized connected computers rather than telephones – has made it possible to carry our whole universe with us, giving us unrivalled means of projection. They have proved extremely useful during periods of confinement, sometimes making the distant more accessible than the near. In this space, which is now also the space of our connection capacities, the unambiguous measurement in meters or miles is no longer appropriate.

The time reference also loses its fixity. This is generally referred to as acceleration (Rosa 2005). We prefer to characterize it in terms of movement and change. It is, indeed, moving images that we see on our screens. New investigative technologies have also enabled us to realize that we do not live in an infinite, unchanging universe, but a universe with history and boundaries. It sheds new light on the thoughts of Lao Tzu and Heraclitus, for example, for whom "it is a new sun every day"[8]. Thus, charged with new aspects and layers, space and time lose the simplicity of their classical ordering.

8 Which is right, even on a physical level, at a certain degree of observation.

The type of links between facts is also at stake: the echo and mirror mechanisms are not characterized by relationship but by interrelation and are at the foundation of the operating modes of search engines, social networks and financial markets alike. In such a context, it is difficult to take as our sole support the necessary, durable and reversible relationships that classical science has chosen to emphasize. The existence of loops, the formation of clusters and the possibility of bifurcations must be taken into account; they call for a place to be given to states and relations that are not just necessary, but also contingent; not just durable, but also non-durable[9]; and not just reversible, but also irreversible.

This highly technical moment ultimately marks the "end of naturality": there are no longer any places, including the seabed, that are in their natural state, that have not been transformed by the work of humans, particularly by our technical tools and practices. We can no longer conceive of reality as being composed of nature on one side and humans on the other side. Instead, reality takes the form of an intertwining of facts, to use the terminology of the early Wittgenstein ("The world is the totality of facts, not of things" (1921)). This intertwining is proving unstable, just as instability has been discovered in the universe. On a global scale, as we are witnessing with environmental issues, it leads to a point where intentionality and human actions transform what we have learned to call nature, and take the ecosystem with it.

The point here is not to make a value judgment on this transition from the infinite universe to the reflexive system. It is a question of how to deal with it and the appropriate concepts to guide us through it.

12.4. The unsuitability of the Enlightenment framework

Faced with this transition, all the components are there for us to perceive the contemporary era in the form of a crisis: extreme events always occur, and they are the events that the media prioritize. We can add to that the widespread misuse of language, whereby anything that deviates from the statistical average is described as "abnormal", when the average is merely a point around which a bundle of facts may – or may not – be ordered. Then, we can clearly see the emergence of three disjointed lines of reference: that

9 See Caye (2020) on the process of duration.

of science, which largely reports facts around average situations; that of the media, which focuses on extreme facts, particularly negative ones, and is much less attentive to progress; and that of our personal experience, which is part of a variable space, not just made up of low and average points. Among these three lines, coherence is impossible.

In today's technology-driven world, it is also no longer possible to distinguish between market and non-market spaces or between public and private spheres in the way that Enlightenment philosophy did.

In the space of half a century, our daily relationship with the commercial sphere has been completely transformed. Today, we get money and pay via screens, giving us access to economic capacities that were previously reserved for the most powerful. E-commerce now makes it possible to consume from anywhere at any time, diluting the economic act throughout life's time and space.

High-powered networked computing has led to an expansion of commercial reality based on the domination of the segment of the economy that has best been able to take advantage of these new capabilities, that is to say, the financial markets. In 50 years, their topography has changed radically. There are no longer a crowd of brokers on a trading floor but a stack of screens that mostly operate automatically, with human intervention in the design and parameterization of these automata. This is more than just a change in appearance; it is a radical transformation of the way markets work.

Since the 1980s, indexed investments have multiplied and derivative products have been deployed (hedging instruments against currency and interest rate risks, etc.). These transactions were legitimized by theories of market efficiency, according to which the best you can do is follow the market[10]. The tools used, the business theories (agency theory in particular) and the amounts involved have all merged to give financial markets a dominant role over the other two types of economic market (goods and services; labor). Today, their "leading edge" is high-frequency trading, which largely abstracts them from the boundaries of space and time[11].

---

10 For further details, please refer to Charolles (2008, pp. 120–138; 2022, pp. 181–196).

11 Orders are decided and carried out at speeds close to the speed of light.

Concomitantly, the separation between public and private spheres has become much harder to establish. Social networking, cell phone capturing capabilities, video surveillance and geolocation mean that the private sphere is constantly under threat of being made public. The preservation of private space was one of the guiding principles of the republican and democratic thinking of the Enlightenment, with the *res publica* protecting the *oïkos* ("household" or "basic unit of society" in ancient Greek) and the *oïkos* in return not hindering the deployment of the *res publica* (Kant 1795). In this movement, the place of politics is more difficult to define.

Our technical advances have also redrawn the boundaries of life and death. In affluent countries, life now appears on screen (in the form of ultrasound), as does death in hospital (flat cardiograms or encephalograms) and, in many cases, illness for both diagnosis and treatment. This obviously opens up new intimate and political questions: should we stop the process of life when a fetus does not appear to have the expected shape? When should a terminally ill patient be taken off a respirator? Habermas deals with these issues by relating what we have to accept as the natural foundations of human life (2001).

The problem, however, is precisely whether it is "the natural foundations of people's self-understanding" (ibid., p. 112) that should be referred to in a manufactured world. Another solution, which is relevant in practical and fruitful in theory, is not to relate law to nature but to a construction. In this case, as far as the human being is concerned, rights are based on a claimed substratum: "human rights", rights that are not natural but rightly declared and constructed. This does not mean that they are reduced to arbitrary agreements. The way in which humans have defined their rights is the result of a long social and political process; it has been built over time with difficulty and has been the outcome of struggles against arbitrariness, domination and despotism. It is hard to see why this substratum would be any less powerful than a reference to "nature" or "natural foundations". This substratum also has the advantage of making it clear that there is no reason why the use of technologies should not be subject to these rights and principles (respect for the private sphere, private correspondence, etc.)[12].

---

12 See Feenberg (1995) and his conclusion to this book on the conception of technologies.

A world in which the easy separation between public and private market and non-market spheres disappears, where intertwining creates complexity, is clearly a post-Kantian world[13].

12.5. A place for politics and the decidable

12.5.1. *The question of frames of thought*

Descartes' *Discourse on Method* successfully answered the enigmas facing the world at the time, enabling freedom to emerge in the political field in a new form: free will. This method can be understood in the same way as harmony in classical music or architecture. Today, we similarly need frames of thought that are consistent with all our works (borrowing from jazz and performance, the arts of displacement, repetition and transgression) and that enable us to account for and orient ourselves in a reality shaped by our technologies. Here are a few steps[14].

The logic of the independence of observations, which could be applied to simple physical, biological or social phenomena, has been called into question on both practical and theoretical levels. The necessary, durable and reversible relationships formalized by classical science can no longer be taken as a standard model. These categories correspond to the major modalities for the register of "things" (for pure physics on a Newtonian scale). As we saw at the beginning of this chapter, the register of facts calls for more open categorizations, ranging from the durable to the non-durable, the necessary to the contingent and the reversible to the irreversible.

Similarly, we have decided to build logic on two statements: true and false. However, in the world of life, or rather in the forms of life[15], events occur in a much less binary way. If it is true that "I have not eaten", it is certainly false that "I have eaten". However, I may not have eaten because I am not hungry, and thus the statement "I have not eaten" will be destined to last, or I may not have eaten because I have not had time to eat yet, so the statement "I have not eaten" will quickly become false. Some false

13 We expanded on this point in Charolles (2008).

14 See Part 4 of *Philosophie de l'écran* (2013) for further explanations.

15 "*Lebensform*", not "*Lebenswelt*". We thank Claude Imbert for drawing our attention to this point. On the concept of logic, the reader can refer to his work, in particular his article in *Intellectica* (Imbert 1987).

statements are intended to last from an instant t to an instant t+1 ("Socrates is dead"), while others are not. In *Philosophie de l'écran*, we developed a truth table constructed in this way (2013, pp. 266–267). It allows for indeterminacy between different temporal states (state t+1 cannot be predicted from state t), while respecting the principle of causality (each event is indeed the result of one or more causalities).

This foundation allows us to envisage new ways, compared with Kant, of bringing mechanical causality and free will into play in the same world. With "the starry heavens above me and the moral law within me" (1966 [1788], p. 173), Kant solved the problem by distinguishing two radically different orders, each unable to influence the other and yet coexisting nonetheless. As we pointed, such a division between the physical and moral eras is not suited to grasping a fabricated world. There are other ways of circumventing the difficulty of making free will and mechanistic causality coexist. Here is a way that comes from examining temporality as we see it operating in the universe on our screens, and in a slightly unexpected manner: on a very long timescale, physics events, like human-induced events, are subject to indeterminacy and randomness (Darwin goes in this same direction for biological organisms). Therefore, even for facts studied by physics, determinism is erroneous (on a certain scale of time and space). For facts resulting from our actions, randomness occurs on a much shorter time scale, which is accelerated by the screen. The reason for this is the intervention of intentionality (to use a more modern word than free will), which is precisely humans' ability to cause hazards and build the world around them (the typical example being the Napoleonic case in France). Through these ideas, we can find a continuum between the universe and the world of life, but in a mode that is not that of absolute determinism and can go as far as randomness (with, of course, varying degrees depending on the types of facts being studied).

A less reductionist stylization of the world (no longer based on a physicalism that predates the intervention of quantum theories and relativity) would not only provide a better account of how reality works; it would also be an opportunity to recognize the place of our individual and collective capacities to act; we do this around the concept of decidability[16].

16 We are not speaking of "decidability" in the computing sense here.

12.5.2. *Decidable support and the role of rules*

The decidable manifests itself daily in the simplest acts of life (eating one thing rather than another) and is also limited by certain conditions (having enough money to buy the food we want, to continue on the same theme). It also exists at the collective level, as expressed in electoral votes, for example, but also in the rules and norms we have shaped over time. It does not mean that we can do everything. However, our capacity to act effectively is greater than we believe, so that we are not reduced to solely relying on the paradigm of safety and the containment of bodies and minds. The decidability of a situation rests upon the technical rules that have been set.

For human activities, some rules (as understood by Wittgenstein) play a decisive role in shaping the world. These rules are the result of processes involving intentionality and are indicators of the construction of our societies. They shape the world of life in its economic, political and social aspects (accounting, statistical and electoral rules; rules for the operation of financial markets; computer coding rules; etc.). They deserve attention.

In the financial sphere, for example, the rule that quotations can be made to two decimal places after the decimal point is fundamental. It was accepted by the SEC (the public authority in charge of stock market regulation in the United States) in the early 2000s (instead of quotations to one sixteenth of a unit, which themselves replaced quotations to one fourth of a unit). It made computerized order processing unavoidable (no broker can remember quotations that precise). This rule is largely responsible for the depth acquired by financial markets (the transaction volumes that take place on them) and for the development of high-frequency trading, which makes the operation of financial markets very tricky to monitor for the authorities responsible for regulating them[17].

While the rules and standards we have set ourselves are not arbitrary and are the result of a long-run process, they are the product of decisions that are not immune to power relationships and that some habits and some frameworks can sustain (the way in which the public sector is treated in national accounting or financial regulations, the choice of the term "artificial intelligence", etc.). This is precisely where the decidable finds a space to be

17 See: https://www.sec.gov/tm/reports-and-publications/special-studies/algo_trading_report_2020.

exercised, where our collective capacity for action can be brought into play effectively.

To illustrate this potential, the network era could provide an opportunity to enhance the value of labor, the neglected origin of digital transformation. If accounting practices established during the Renaissance did not include labor as a value in company balance sheets, it is because wage-earning practically did not exist for the Florentine bankers and Venetian long-distance traders by whom modern double-entry bookkeeping was invented. The bias toward a single factor of production (capital) as the basis for corporate value was not challenged during the first industrial revolutions, even though Adam Smith made labor the original source of all the wealth of nations in his eponymous book. Today, the process of innovation and intangible creation is at the heart of digital tool deployment and permeates all economic sectors. It is no longer a question of using the dexterity and care brought to work by employees, like in *The Wealth of Nations*, but of using all their qualities (intelligence, interpersonal skills, etc.).

However, instead of relating the resulting creations to employees, we see the curious term "artificial intelligence"[18] being imposed, suggesting that today's machines are self-constructing. The same illusion thrives around "algorithmic reason". Although today's machines are much smaller than yesterday's machines, they are not self-generating and still rely on human ingenuity. Even so-called *deep learning* systems need a human to specify the basic building blocks to get started. As for the "intelligent" nature of these computers, as conceived by Turing in his 1950 article, it is difficult to neglect Searle's Chinese room rebuttal (1980): reproducing or even producing is not equivalent to communicating and understanding. This is part of a general mindset in which the contribution of work to the economic sphere is not recognized, nor the fact that technical creations are its outcome. Accounting standards still overlook this point[19].

Let us us give two final examples of the potential offered by the notion of the decidable as applied to social, economic and political rules and norms.

---

18 See Chapter 5 of this book by Jean Lassègue on the inappropriateness of the term developed in 1956 and Chapter 6 by Éric Guichard on the fact that technologies are part of culture.

19 This is not the place to go into further detail on this subject. Those interested can refer in particular to our previous work (Charolles 2006, pp. 23–36 and pp. 183–204; 2021, pp. 25–38 and pp. 2013–234) and to a communication from 2019.

The first concerns the definition of emergency and its consequences in the political field. As mentioned in the introduction, states of emergency have been on the rise since September 11 (at least), the result of which is limited freedoms and security being deployed using all available technologies. However, there is no democratically accepted definition of what constitutes an emergency justifying a state of emergency, in other words, specific rules and/or the mobilization of human or financial resources. Thus, the state of ecological emergency, which has been evident at least since COP 21, does not exist in legal terms, unlike what happened with terrorist attacks, the financial crisis in 2008 and 2009 and the Covid-19 pandemic. The reduction of global inequalities and the development of democracy have not been elevated to the status of legal emergencies, nor have they benefited from measures of comparable magnitude. In addition, the measures adopted under states of emergency have tended to be based on exceptional procedures and measures, rather than on the exceptional mobilization of pre-existing common-law mechanisms. In France, for example, the management of Covid-19 was based on the government's use of the power of parliament, and new legislation was introduced to monitor the isolation of sick people, even though the legal framework existed already.

The concept of the decidable can provide a basis for dealing with these issues in a less discretionary way; that is, to limit situations where we have to declare an emergency, reduce the range of derogatory measures taken under emergency conditions and use democratic procedures of ordinary law as far as possible. The appropriation of technologies (insofar as they are our creations) is part of this process.

The second example concerns the configuration of technical tools used by companies and all forms of organization, including governments, in routine or emergency situations. As soon as we are clear about the fact that they involve decisions, we can legitimately expect that they respect certain conditions. In this regard, we propose a two-pronged approach: epistemological and ontological. It consists of a double questioning regarding both the solidity and robustness of the technologies used as well as the conception of the subject and its intentionality on which they are based. When faced with a complex, interdependent and uncertain world, this can provide a way of avoiding the technicist illusion and of finding balance and equilibrium. Without going into too much detail, we would like to point out that this epistemological and ontological questioning could challenge and give substance to the now widely promoted ethics of technologies.

12.6. Conclusion

In this chapter, we have shown how the development of technologies places us in a new universe of reference, which we called a reflexive system. We stressed that if we stick to a human/nature opposition that leaves no adequate room for technologies, we will not be able to find a suitable conceptual foundation. To provide a framework for thinking about this world, its hazards and the ways in which decisions are made, we have to put forward the concept of the decidable at both an individual and collective level.

We have stressed that the concept of the decidable has no reason to be reduced to an illusory dream of philosophers and that technologies are the result of human work and a part of culture – a point that is echoed throughout this book. These are works over which we have control, particularly in the construction of the rules that govern their deployment. We are often unaware of this and do not have the opportunity to assert it, as we are "entangled"[20] in rules that we have not sufficiently reviewed or whose scope we simply ignore. This shows how important philosophical questioning is regarding the uses and practices of technologies and the need for a philosophical moment.

12.7. References

Agamben, G. (2005). *State of Exception*. University of Chicago Press, Chicago.

Biondi, C. (2022). Une relecture des enjeux politiques du numérique. PhD Thesis, Sciences Po and Columbia University.

Braudel, F. (1981). *Civilization and Capitalism. 15th–18th Century*. Collins, London.

Caye, P. (2020). *Durer. Éléments pour la transformation du système productif*. Les Belles Lettres, Paris.

20 See Wittgenstein (1967, p. 50) on the entanglement in our own rules as a philosophical issue.

Chardel, P.-A. (ed.) (2014). *Politiques sécuritaires et surveillance numérique*. CNRS, Paris.

Charolles, V. (2008). *Et si les chiffres ne disaient pas toute la vérité ?* Fayard, Paris.

Charolles, V. (2013). *Philosophie de l'écran*. Fayard, Paris.

Charolles, V. (2021 [2006]). *Le libéralisme contre le capitalisme*. Folio, Paris.

Charolles, V. (2022). *Se libérer de la domination des chiffres*. Fayard, Paris.

Debord, G. (1970). *The Society of the Spectacle*. Black and Red, Detroit.

Descartes, R. (2001 [1637]). *Discours de la méthode*. In *Œuvres philosophiques, vol. I*. Garnier, Paris.

Dupuy, J.-P. (2002). *Pour un catastrophisme éclairé : quand l'impossible est certain*. Le Seuil, Paris.

Dupuy, J.-P. (2012). The precautionary principle and enlightened doomsaying. *Revue de métaphysique et de morale*, 76, 577–592.

Feenberg, A. (1995). *Alternative Modernity: The Technical Turn in Philosophy and Social Theory*. University of California Press, Berkeley.

Foucault, M. (1970). *The Order of Things: An Archaeology of the Human Sciences*. Tavistock, London.

Foucault, M. (1975). *Discipline and Punish: The Birth of the Prison*. Penguin Books, London.

Guichard, E. (2017). La philosophie des techniques revue à l'aune de l'internet et du numérique. In *Le numérique, enjeux et débats*, Chazal, G. (ed.). Presses Universitaires de Dijon, Dijon.

Habermas, J. (1968). *Technik und Wissenschaft als "Ideologie"*. Surkamp, Frankurt.

Habermas, J. (2001). *Die Zukunft der menschlischen Natur. Auf dem Weg zu einer lilberalen Eugnik?* Surkamp, Frankurt.

Husserl, E. (1954). *Die Krisis der Europäischen Wissenschaften und die Transzendentale Phänomenologie*. Husserl Archiv, Den Haag.

Imbert, C. (1987). Entre logique naturelle et intelligence artificielle. *Intellectica*, 4, 7–40.

Kant, I. (1788). *Kritik der praktischen Vernunft*. Hartknoch, Riga.

Kant, I. (1795). *Zum ewigen Frieden. Ein philosophischer Entwurf*. Friedrich Nicolovius, Königsberg.

Koyré, A. (1957). *From the Closed World to the Infinite Universe*. John Hopkins Press, Baltimore.

Mandelbrot, B. and Hudson, R.L. (2004). *The (Mis)Behavior of Markets. A Fractal View of Risk, Ruin and Reward*. Basic Books, New York.

Marcuse, H. (1964). *One-Dimensional Man: Studies in the Ideology of Advanced Industrial Society*. Beacon, Boston.

Quine, W. (1960). *Word and Object*. MIT Press, Cambridge.

Rosa, H. (2005). *Beschleunigung. Die Veränderung der Zeitstrukturen in der Moderne*. Suhrkamp, Frankfurt.

Searle, J.R. (1980). Minds, brains and programs. *The Behavioral and Brain Sciences*. 3(3), 417–424.

Smith, A. (1776). *An Inquiry into the Nature and Causes of the Wealth of Nations*. Strahan and Cadell, London.

Turing, A.M. (1950). Computing machinery and intelligence. *Mind*, 49, 433–460.

Wittgenstein, L. (1921). Logisch-philosophische Abhandlung. *Annalen der Naturphilosophie*, 14, 185–262.

Wittgenstein, L. (1967). *Philosophical Investigations*. Basil Blackwell, Oxford.

Conclusion

Marcuse's Critique of Technology Today

C.1. Introduction[1]

I start with the question, why Marcuse? I would not have to ask that question for Adorno's critique of technology or Deleuze's critique of technology, supposing they had one. These are the fashionable thinkers who everybody is sure have something important to tell us. Marcuse has been eclipsed by them in the last 30 years or so. However, there was a time when he was the most famous philosopher on Earth with the possible exception of Sartre. Something happened and I think this is partly connected to his very radical critique of science and technology.

Under the influence of Habermas, critical theory dropped the critique of "instrumental rationality", especially in the form it took in Marcuse's work. The new field of science and technology studies rejected his critique along with that of Heidegger and other critical predecessors as technophobic. Yet Marcuse addressed issues involving science and technology that are even more pressing today than when he developed his approach. The main lines of his radical stance are validated by the multiple crises we face.

Habermas rejected the "secret hope" of the early Frankfurt school for the re-enchantment of nature in a socialist society. For him and his many followers, technology is just a practical convenience that has no capitalist

Conclusion written by Andrew FEENBERG.

1 A premiliary version of this text was offered in French to the conference "Pratiques et usages contemporains des philosophies des techniques" at the College international de philosophie, June 2021.

bias, no relation to the problem of domination, unless it intrudes into the lifeworld, that is, the communicative sphere. We can agree that technified communication poses problems, but this very limited Habermasian approach does not make much sense in a world concerned with many other technical issues, climate change, the pandemic and surveillance.

Something similar happened with science and technology studies (STS). In the 1960s, a number of thinkers, including Marcuse, addressed Heidegger's famous "question concerning technology". These thinkers raised broad concerns about technology and its dangers. However, when STS became the dominant field in the study of technology, these concerns disappeared. STS evolved out of a conscious decision to make what Wiebe Bijker called an "academic detour", in other words, to stop talking about nuclear power, pesticides and the dangers of technology generally in order to treat it as just another academic subject. This enabled the new field to establish itself in the university but without sociological self-awareness. In fact, its success was largely due to increasing protests over the consequences of science and technology. Ironically, STS was boosted by the very concerns it chose to ignore.

By the 1990s, the issues Marcuse raised in the 1960s appear as a mere historical background to contemporary critical theory and science and technology studies. These fields did not do the "critique of technology" anymore. The environmental movement has changed this. Younger scholars now approach politically sensitive issues in science and technology. However, capitalism is rarely mentioned and the very concept of "critique" is rejected by some as outmoded and impotent.

C.2. The role of capitalism in the social construction of technology

It is strange that these very sophisticated theoretical traditions are reluctant to talk about the role of capitalism in the social construction of technology. Everyone else seems aware of it. Science and technology have evolved in a social environment from which they have accumulated various biases that arise from the priorities of capitalism. Now that it is time to overcome these biases, scientists and technologists responding to public concerns face obstacles both from their tradition and from their corporate and government masters. Surely, this is an occasion for theory.

As he was a Marxist, Marcuse entered the debates over the question of technology with a strong sense of the role of capitalism. When the environmental movement took off in the late 1970s, Marcuse responded to it, at a time when many Marxists dismissed it as a distraction from class struggle. He could see in the environmental movement exactly what he had been talking about all this time, the conflict between human life and capitalism. Marcuse thought that this conflict would someday be resolved by socialist revolution, an event we are even less inclined to wait for than he was. In the meantime, he supported the political struggles over technology inspired by the early environmental movement.

The emphasis on the role of capitalism goes against the grain in most philosophical discussions of technology. There is an awful lot of talk about the nature/culture divide, which is supposed to be the cause of all our problems. Or maybe it is Descartes, but it is something like that, some intellectual concept that we now know does not make sense. In reality, the nature/culture divide plays no role in the political fight over climate change. That fight is a conflict between various institutions and publics with different powers and priorities. The most powerful of these institutions are certain capitalist corporations that are simply interested in making money. Nor are the people who resist, who demonstrate, the kids who skip school on Friday, involved in this abstruse philosophical debate over the relation of nature to culture. They are concerned about their future, which depends on what we do about the climate.

C.2.1. *The importance of the* Lebenswelt, *the lifeworld*

There is a second reason Marcuse anticipates our contemporary concerns. Although a Marxist, he turned to phenomenology to deal with the inability of traditional Marxism to understand the nature of resistance to capitalism. The Marxist idea of class consciousness is excessively rationalistic. Resistance is not just about political opinions or having learned something from reading *Capital* or even having economic interests that would be better fulfilled under socialism.

Marcuse's understanding of resistance was influenced by his studies with Husserl and Heidegger. He bridges what were the two major philosophical traditions of European philosophy throughout much of the 20th century. For several generations, philosophers were concerned with the contest between

Marxism and phenomenology and existentialism. Marcuse crossed the lines between these adversaries in the late 1920s and early 1930s. At first, he referred openly to phenomenology but once he joined the Frankfurt school, these references were muted. He played down this phenomenological past and yet it is still there and explains some of the stranger and more interesting things in his critique of science and technology.

He learned from phenomenology the importance of the *Lebenswelt*, the lifeworld, that is, the first person relation to experience. This concept is essential for any philosophy that wants to escape from scientific naturalism, that wants to be able to credit values as something real and significant with which people are engaged as both thinkers and actors. In lived experience, reactions to the problems of society have an existential weight that is missing in a lot of Marxist theory. This shows up most significantly in his understanding of the new left. Its relevance today is evident in struggles over race, gender and the environment.

C.2.2. *Technology as a one-dimensional framing*

Marcuse proposed three related critiques of science and technology. They are all interesting and they get us on the right track toward a solution to the "question concerning technology", although we will need to go beyond his account in important ways. The first two critiques are summarized in the phrase "technology is ideology". This phrase is confusing because we usually think of ideology in contrast with science, the one false, the other true. However, technology is ideology for Marcuse for entirely different reasons. He means that technology is not a neutral instrument but rather it shapes the things that we think and do and limits thought and action to what the technified social world has to offer.

Marcuse's first critique argues that this is a new kind of legitimation process, different from the role of the bourgeois ideology or the church. It is the actual material world itself that now legitimates the system by prescribing all the needs and hopes that people can entertain. Technology frames a whole way of life and it legitimates that way of life in a modern society concerned with the effective mastery of natural forces.

This leads to the triumph of what Marcuse calls “one-dimensional thinking”. Technology is not only the material culture of the society but also a mode of thought for which all problems are technical problems and have technical solutions. Technical problem-solving within the given framework replaces normative political concerns that might lead to fundamental social change. Where technical thinking takes over, utopian ideas no longer have any currency.

Marcuse’s second critique was presented in an essay on Max Weber’s theory of rationalization. By rationalization, Weber means something very specific: calculation and control of the institutions, calculation of resources and costs of benefits and control of the human elements and material processes. Marcuse argued that Weber’s concept of rationalization was uncritical: calculation and control by whom, calculation and control of whom? Those are the questions that Weber did not adequately address. Under capitalism, calculation and control are handed over to people exterior to the work process, capitalists and their managers. Work is organized from above and not by the common agreement of the people who do the work.

Management from above has consequences for the nature of work and for the kind of things that are produced. For example, factory management typically emphasizes profitability, while workers have other concerns such as job security and health and safety on the job. Control from above also biases technical designs intended to pace work, surveil workers and control the labor force. The assembly line is an example of such a biased technological design. Presumably workers, if they had control, would not employ it. This pattern of control from above extends into every aspect of social life in advanced capitalism and is accepted as the inevitable price of progress.

Weber was unable to criticize those consequences of capitalism because he assumed that control from above was essential to effective collective action in any modern society. His rationalization thesis thus contains, on the one hand, a realistic view of how things actually work under capitalism, but, on the other hand, it masks the bias of capitalism toward domination. There is almost always an alternative in which control would not come from above but from the organized workgroup itself. Yet Weber does not take this into account but simply assumes that rationalization is controlled from above.

C.2.3. *Ontological critique*

Marcuse extended this second critique to technology and technical reason as such under capitalism. The argument is further developed and given a much larger context in *One-Dimensional Man*. There Marcuse proposes a more complex ontological critique that encompasses and frames these social critiques.

By ontology I mean an account of what is real. Marcuse asks what is real in a society dominated by science and technology, and he follows the argument of Husserl and Heidegger according to which natural science defines reality in terms of the primary qualities of matter in motion, measurable aspects of the material world. Marcuse argues that this results in a one-dimensional form of experience, one-dimensional in the sense that other dimensions of experience that are available to everyday consciousness are devalued. Science does not need your dreams and hopes, and it does not need your morality and sense of beauty. It has its own austere normative claims which do not intersect with our daily concerns. Despite the revolution in physics that led to relativity theory and quantum mechanics, nothing has changed at this very general level.

Although he was seen as anti-scientific in his time, Marcuse's argument is not with physics as such, but rather with its interpretation as a universal ontology. Science constructs an object, the world as matter in motion, that is adjusted to its mathematical and experimental methods. When that object is taken for reality as such, everyday experience is limited to those same parameters. Marcuse uses the term "one-dimensional" to describe the kind of thinking that results from the intrusion of what are properly scientific ideas about the nature of reality into everyday experience. He explains this in terms of a unique synthesis of phenomenology and Marxism.

Husserl argued that science arises from the lifeworld, the everyday world of ordinary experience. Science picks up concepts from the lifeworld and refines them into instruments of scientific knowledge. We have, for example, an experience of space and ways of measuring space in our everyday experience. These experiences and measures are refined in the physical concept of geometrical space.

The notion that science emerges out of the lifeworld suggests to Marcuse a consideration that Husserl would never have entertained. From his Marxist point of view, he argues that the lifeworld changes with changes in the mode of production. The lifeworld is shaped by the fundamental practices and concepts that emerge from the capitalist structure of experience and then is brought into science, which thus depends ultimately on capitalism. Modern science reflects the nature of experience under capitalism.

Marcuse amplifies this claim in terms of Heidegger's late philosophy of technology. The world appears in Heidegger as a system of resources available for domination by human subjects. Heidegger's term for the project of domination is "*Ge-stell*", the framework that frames the world. The word *Gestell* in German can be used to signify a skeleton: there is a scientific skeleton to everything in a technified society.

There is reciprocity between Husserl's theory of science arising from the lifeworld and Heidegger's theory of technology as domination. Qualities derived from the lifeworld and valued in science, such as the measurable aspects of reality, become practical in everyday experience where they are used to dominate nature. Meanwhile, other qualities that science ignores – Marcuse is particularly concerned with beauty – are excluded from the considerations determining technology.

Heidegger shows that modern technology differs from ancient technology because it views the world as fungible resources rather than as a source of potentialities that could be actualized with help from human craftsmen. In this, Heidegger relies on Aristotle's teleological concept of potentiality. In Aristotle's view, the craftsman brings out the potentialities of the materials just as the farmer brings out the potentialities of the soil and plants. For Aristotle, potentiality is a real dimension of being even though it is latent rather than actually present. He concludes that a thing can be and yet not be and that it cannot be and yet be. This is his way of qualifying the ontological status of potentialities, which are not yet but which are real all the same.

Potentiality in this sense is a familiar concept. We learn, children grow, forests flourish and so on. We see nature and other human beings in terms not only of what they are in the moment, not only in terms of what we could measure, but also in terms of what they can become and this perception is normatively informed. Science eliminates such "secondary qualities" in order to get to what it can use as facts and data in its work. When the ontology of science penetrates into experience, potentialities are eliminated not just from theory but also from the objects of experience.

What happens when we see experience largely in terms of such a quasi-scientific ontology? We no longer interact with objective potentialities of human beings and nature, those potentialities Aristotle took for real. Once his teleological metaphysics is rejected, what he thought of as potentialities of the things themselves, we take for subjective values. These values are the products of our psychology, our background and culture, and have no ontological status.

Now, to be sure, potentialities as Aristotle understood them were also inscribed in his Greek culture just as are our values. Nevertheless, the notion that nature's potentialities are real, that is, nature is a subject in its own right, could serve as a guide to human action. When science becomes the dominant ideology of society, nature no longer has this independent status. Marcuse argues that if nature cannot guide human action, the only effective values will be determined by the dominant powers, that is, the corporations and the government agencies that act for them. The potentialities eliminated from science will no longer influence the actions of those powers, with consequences such as we see today in the destruction of the environment.

In modern times, the world appears stripped of its own sources of development, available for any kind of purpose that human beings want to impose upon it. Marcuse explains this by reference to Sartre's idea of a "project", which again is drawn from Heidegger. A project in Sartre's sense is not a particular plan but rather it is a scene on which plans can be devised and implemented. Capitalism is a civilizational project in this sense and defines a specific lifeworld. Its project reduces the world to raw materials. In the capitalist lifeworld, things are simply there to be exploited and dominated. Scientific concepts emerge out of this world and reflect it, as do technological designs.

This leads Marcuse to a dystopian conclusion characteristic of the 1960s. In that decade, *Brave New World* and *1984* inspired popular culture. The level of technification was so developed that individuals could imagine everything being brought under the control of the social machinery. This dystopian outlook is renewed now with surveillance on the Internet. Today, "they" know everything about you and so can manipulate you as they wish. In our version of a brave new world, everything is predictable and human beings are just relays between social organizations operated by the dystopian system.

C.2.4. *Dystopia*

By the late 1960s, dystopian popular culture became influential not just with a few intellectuals, as in *Brave New World*, but was commonplace, propagated in films and books that portrayed society as already completely controlled by an elite or by machines. Several popular European films of the 1960s, such as *Fahrenheit 451, Alphaville* and *The Red Desert*, prepared this turn. In *La Dolce Vita,* for example, the novelist hero gives up literature to become an adman and the one character who represents traditional culture ("We can no longer hear these voices") commits suicide because he cannot live in the society emerging around him. Everything is absorbed by the system or destroyed. These films depict a completely technified society.

In *One-Dimensional Man*, Marcuse proposed a quasi-dystopian theory and yet at the same time he held out hope that perhaps forces may exist capable of breaking through the system. In fact, the idea that we are living in a dystopian world captured the imagination of so many young people that it motivated political resistance. The new left emerged in part out of dystopian anxieties. There is a paradox here since the discourse of dystopia generates the resistance to dystopia.

In the 1960s and 1970s, Marcuse developed new ideas around the emergence of these new forms of resistance. The new left had the existential quality he associated with the real possibility of change, the project of socialism. He was never an optimist, but critical thought had once again found a mass base. The proletarian revolutions of the 19th and early 20th century were over in Marcuse's view but the new left brought the socialist idea back to life.

C.3. A "successor science" and "successor technical disciplines"

Marcuse was not content with simply criticizing science and technology. As a Marxist, he also had ideas about the future. He imagined socialism as a different civilizational project from capitalism that would produce a different science and technology. He argued that technology and science would change under socialism because they had become the fundamental infrastructure of the technified capitalist system. What we might call a "successor science" would emerge along with an entirely new organization of the social system. Under socialism scientists will have different experiences as their lifeworld changes and that will result in the development of different scientific concepts. The process of change could take generations, but eventually scientists themselves will create a new kind of science.

What would be different about this new science? This is where things get complicated. The new science will *not* be a qualitative science. There is no way back to Aristotle. The new science will still employ quantitative methods based on precise measurement and empirical facts. It will retain the "rigidity" of scientific method as we know it; however, it will also incorporate potentialities into its structure. Art is the domain in which potentiality can be freely imagined. If art and science could merge, then science would no longer be confined to the one-dimensional world of primary qualities but could incorporate secondary qualities.

C.3.1. *Escaping the ontological paradigm of physics*

This is where I have my doubts about Marcuse's argument. There is no way for a formal science such as physics to incorporate the secondary qualities from which it is abstracted. Modern science is based on equations, unlike earlier sciences which were based on generalizations and analogies. Equations have no room for concrete experience.

Consider a simple equation such as distance equals velocity times time. Distance traveled is a function of the speed and time of travel. How can you ever get anything concrete into a scientific formula like this? All you can do is specify the variables relevant to the equation, for example, the measured distance between your home and workplace. This distance cannot include the aesthetic qualities of the buildings encountered along the way. The formula repels such concrete qualities and sticks rigorously to its abstraction. There is simply no way to incorporate concrete data back into an equation, but equations are what science is all about.

Marcuse wants a different ontology to emerge with a different social world. However, he does not have an explanation for how that can happen given his focus on physics as the ontological foundation of modernity.

Why would Marcuse have made this assumption? The social context suggests an explanation. In the early 1960s, science and technology were idealized in the United States to an unusual degree, very different from today. Everything seemed to depend on science, confirming the arguments of Husserl and Heidegger for whom physics was the basis of the modern ontology. Only in the late 1970s did an environmental movement strong enough to have broad public impact appear. By then, Marcuse was nearing the end of his life. He argued that environmentalism strives to protect potentialities of nature favorable to human life that were being destroyed by industry, but he did not go back and revise his earlier theories of science and technology in response to this new situation.

We need to escape from the assumption that physics is the ontological paradigm in modern society. Only if we abandon that assumption can we find a different way for reason to operate in a different kind of social system. Nevertheless, despite these limitations of Marcuse's argument, he introduced several useful concepts that we can apply in a way he did not anticipate but which I think he would have approved. I have developed these concepts in my forthcoming book *The Ruthless Critique of Everything Existing: Nature and Revolution in Marcuse's Philosophy of Praxis* (Feenberg 2023). I will offer some comments on this alternative at the end of this text.

C.3.2. *Translation and object transformation as a politics of technology*

In *One-Dimensional Man*, Marcuse uses the term "translation" to explain how the new science he imagines will "materialize" values. When I noticed this recently, I was surprised. A similar concept was developed in science and technology studies 30 years later. Latour, for example, describes the translation of norms in technical design as "delegation". He uses the example of a speed bump in the road. It translates or delegates the normative principle "go slow" into the agency of the material world. This way of thinking became obvious to everyone studying science and technology by the 1990s, but the concept was there in Marcuse in 1964, derived in part from the work of Gilbert Simondon. However, he did not quite figure out all its implications. His account of the translation of values into technical terms sounds a lot like economic planning.

Marcuse developed another useful concept in his critique of Weber and in *One-Dimensional Man*. That concept is object construction. The objects of the sciences were traditionally viewed as inherent in nature but Marcuse argues that these objects are historically contingent. He thus historicizes Husserl's claim that the lifeworld is the source of scientific concepts. When the lifeworld changes, the object of science changes as well. Marcuse concludes *One-Dimensional Man* by trying to define the new object of a successor science under socialism.

Unlike in the 1960s when Marcuse developed his theory, there is now a great deal of contestation of science and technology. They are no longer sacrosanct. Science and technology are subject to much criticism, some justified and some unjustified, and this criticism issues in political struggles. For example, the feminist movement resisted traditional childbirth procedures and demanded changes in the treatment of women. The movement of gay people around AIDS resisted the established methods for experimental medicine. Various environmental movements address concerns about air and water pollution, toxic waste dumps and, eventually of course, the biggest of them all, climate change. We are now used to social movements altering and changing the scientific-technical disciplines that actually determine social life. Studying the consequences of these struggles gives us a very different paradigm of science from Marcuse's view in the 1960s.

The notion that there is interaction between the historically changing lifeworld and the sciences is fundamental to Marcuse's thought, and it is fundamental to our own situation in ways he did not anticipate. His concepts of object construction and translation can be applied to all the technical disciplines that structure our world, the management sciences that administer us, the food science that determines our diet, the civil engineering which builds bridges and automobiles, the medicine that takes care of our physical health, the architecture which constructs spaces and determines their sustainability and so on. We need reforms in all these fields. A changed lifeworld makes new demands on science and technology. This can be conceived as ontological change in the sense that the objects of the disciplines change in function of changes in the social world.

Today social movements alter the objects of the socially embedded scientific-technical disciplines and the designs of the technologies they support through the translation of lifeworld issues and concerns. These disciplines and designs organize modern societies, not natural sciences such as physics which are only servants of the effective fields that do the actual work. The translation connects those lifeworld concerns with the actual construction of the environment.

These successor technical disciplines are not as radically different as Marcuse's imagined new physics, but they improve the quality of life in a technified society. Most importantly, the emergence of political struggle over science and technology promises to keep open the future.

I have been working on this question in recent years, informed by these political struggles. In my book *The Ruthless Critique of Everything Existing: Nature and Revolution in Marcuse's Philosophy of Praxis*, I follow Marcuse in arguing that the lifeworld and sciences interact conceptually and practically, but I add that this gives rise to an ongoing politics of technology. That politics is unique because it crosses the gap between the lifeworld and the disciplines in both directions. The lifeworld is informed by scientific concepts that grant experience meaning and technical disciplines respond to public concerns by modifying their object.

For example, public concern about the health effects of air pollution arose from both personal observation and medical opinion widely shared with the public. This concern eventually reached politicians and regulators, who obliged automotive engineers to modify the automobile to take account of air quality. The automobile started out as a means of transportation and a status object, but it has become a new kind of environmental object as well, both for consumers and engineers. In order to transform the automobile into this new kind of object, the engineers had to add chapters to their textbooks on catalytic converters and fuel injection and so on.

In the case of climate change, we understand what is happening as the result of carbon dioxide in the atmosphere. Our conception of the climate is influenced by science, but science is also influenced by pressures from the lifeworld, pressures to mitigate these effects. In both cases, the object "climate" has been transformed. For us, it has taken on properties known to science while also continuing to concern us as a future environment. For science, the climate is no longer simply an object of observation and prediction but has become a manipulable object of technical control.

C.3.3. *Communicating rationalities*

This is a pattern that appears repeatedly as environmental concerns become more and more significant. The pattern is also repeated in other fields such as medicine where public concerns affect scientific ideas about how to do things. Struggles over science and technology have gradually reconfigured the public sphere. This is an enormous change Marcuse could not have anticipated, although it is consistent with his emphasis on the role of science and technology in society.

To be sure, there are dangers in this new public sphere, as the anti-vax movement demonstrates. However, in the last 75 years public involvement has brought huge beneficial changes in medicine and the environment. These changes are not as visible as the anti-vax movement because successful social movements are quickly forgotten. Today, we take it for granted that clean air is important, but not so long ago automobile manufacturers were busy spending millions of dollars lobbying governments to prevent "expensive and unnecessary" pollution controls from being introduced.

Rational though it is, environmental protection is not obvious but only becomes so in the course of history. The history of the presence of exit signs in theaters illustrates this strange dependence of rational arrangements on historical incident. There is an obvious reason for the presence of these signs: the audience needs to get to the exit in an emergency. However, that reason was not sufficient to get exit signs in theaters in the United States before 1903. In 1903, there was a fire in the Iroquois's theater in Chicago and hundreds of people died because they could not find the exits. After that all the city councils in the country demanded that every theater have exit signs.

Clearly, it is rational to have exit signs, but rationality did not get the signs up until a contingent historical event placed them on the agenda. We are in the same situation with climate change. Of course, it is rational to mitigate it but that is not happening because it is rational. It is happening because social movements and changes in public opinion put on the pressure to make the people in charge take seriously this particular rational consideration among the thousands of others they could worry about.

There is a fundamental difference between the view that I have developed here on the basis of Marcuse's work and his own theory. He believed that scientific change would occur *after* a socialist revolution, which would modify the structure of the lifeworld to bring back a sense of potentialities lost in the dystopia of one-dimensionality. In fact, there is no such dystopia. The lifeworld has retained a sense of potentiality to a greater degree than Marcuse believed likely or possible. It continues to motivate struggles to preserve and protect both human beings and nature.

At the time Marcuse elaborated his theory, the legitimating power of technology appeared to be uncontested and uncontestable. Indeed, it was very unusual for the public to engage with the science and technologies that affected their lives. However, more recently, people have had the chance to speak up in a way that was not the case before. This is a democratic advance. The system is now much more vulnerable to public pressures than ever in the 19th and early 20th century. This is particularly important where climate change is concerned.

Where this will lead, we do not know, but there is precedent for our situation in the labor movement. Generations of workers struggled against capitalism long before there was any plausibility to the notion of socialist revolution. They did not need a final goal of some sort to identify problems and injustices and solve them. After the collapse of attempts to reform the Soviet Union, we too have no clear sense of where our struggles might lead. Will they culminate in a new form of socialism as Marcuse hoped? In the meantime, we will struggle to modify the technical system, and this is something that we have the power to do to a surprising degree.

C.4. References

Feenberg, A. (2023). *The Ruthless Critique of Everything Existing: Nature and Revolution in Marcuse's Philosophy of Praxis*. Verso, London.

List of Authors

Bruno BACHIMONT
CosTech
UTC
Compiègne
France

Pierre CAYE
Centre Jean Pépin, CNRS
ENS
Paris
France

Pierre-Antoine CHARDEL
IMTBS
Évry
and
LAP, CNRS
EHESS
Paris
France

Valérie CHAROLLES
LAP, CNRS
EHESS
Paris
France

Andrew FEENBERG
Simon Fraser University
Vancouver
Canada

Xavier GUCHET
CosTech
UTC
Compiègne
France

Éric GUICHARD
ENSSIB
and
Triangle, CNRS
ENS
Lyon
France

Élise LAMY-RESTED
SASPRO2
Institute of Philosophy of the
Slovak Academy of Sciences
Bratislava
Slovakia

Jean LASSÈGUE
LIAS, CNRS
EHESS
Paris
France

Anna LONGO
Collège international de
philosophie
Paris
France

Vincent PUIG
IRI
Centre Pompidou
Paris
France

Bernard REBER
CEVIPOF, CNRS
Sciences Po
Paris
France

Tyler REIGELUTH
ETHICS
Université Catholique de Lille
France

Index

A, B, C

D, E, F

G, H, I, J

K, L, M

O, P

R, S

T, W

Other titles from

iSTE

in

Interdisciplinarity, Science and Humanities

2023

GIRAULT Yves
Contemporary Museums: Tension Between Universalist and Communitarian Approaches

MAÎTRE Henri
Aesthetics in Digital Photography

REBER Bernard
Responsible Deliberation, between Conversation and Consideration: Conditions for a Great Democratic Debate
(Innovation and Responsibility Set – Volume 10)

2022

DRATWA Jim
Ethics of Transitions: What World Do We Want to Live in Together?
(Innovation and Responsibility Set – Volume 8)

GIANNELOS Kalli, REBER Bernard, DOORN Neelke
Responsive Ethics and Participation: Science, Technology and Democracy
(Innovation and Responsibility Set – Volume 9)

ROBERT Sandrine
Resilience: Persistence and Change in Landscape Forms

2021

GUCHET Xavier
Care in Technology
(Innovation and Responsibility Set – Volume 7)

2020

ATCHOUA Julien, BOGUL Jean-Jacques, DIALLO Saikou
Digital Technologies and African Societies: Challenges and Opportunities
(Africa Digital Age Set – Volume 1)

CHERICI Céline
From Clouds to the Brain: The Movement of Electricity in Medical Science

GUILLAUME Astrid, KURTS-WÖSTE Lia
Making Sense, Making Science

SCHEPS Ruth, MAUREL Marie-Christine
Mathematics in the Visual Arts

PAVIE Xavier
Critical Philosophy of Innovation and the Innovator
(Innovation and Responsibility Set – Volume 6)

2019

ANCORI Bernard
The Carousel of Time: Theory of Knowledge and Acceleration of Time

BOULEAU Gabrielle
Politicization of Ecological Issues: From Environmental Forms to Environmental Motives

GAYON Jean†, interviewed by PETIT Victor
Knowledge of Life Today: Conversations on Biology

KHATCHATOUROV Armen, with the collaboration of CHARDEL Pierre-Antoine, FEENBERG Andrew, PÉRIÈS Gabriel
Digital Identities in Tension: Between Autonomy and Control
(Innovation and Responsibility Set – Volume 5)

LENOIR Virgil Cristian
Ethically Structured Processes
(Innovation and Responsibility Set – Volume 4)

NIKOLOVA Blagovesta
The RRI Challenge: Responsibilization in a State of Tension with Market Regulation
(Innovation and Responsibility Set – Volume 3)

ZEMBRI-MARY Geneviève
Project Risks: Actions Around Uncertainty in Urban Planning and Infrastructure Development

2018

LOISY Catherine, COULET Jean-Claude
Competence and Program-based Approach in Training: Tools for Developing Responsible Activities
(Innovation and Responsibility Set – Volume 1)

PANSERA Mario, OWEN Richard
Innovation and Development: The Politics at the Bottom of the Pyramid
(Innovation and Responsibility Set – Volume 2)

2017

MAESSCHALCK Marc
Reflexive Governance for Research and Innovative Knowledge
(Responsible Research and Innovation Set – Volume 6)

PELLÉ Sophie
Business, Innovation and Responsibility
(Responsible Research and Innovation Set – Volume 7)

2016

GIANNI Robert
Responsibility and Freedom: The Ethical Realm of RRI
(Responsible Research and Innovation Set – Volume 2)

GRUNWALD Armin
The Hermeneutic Side of Responsible Research and Innovation
(Responsible Research and Innovation Set – Volume 5)

LENOIR Virgil Cristian
Ethical Efficiency: Responsibility and Contingency
(Responsible Research and Innovation Set – Volume 1)

PELLÉ Sophie, REBER Bernard
From Ethical Review to Responsible Research and Innovation
(Responsible Research and Innovation Set – Volume 3)

REBER Bernard
Precautionary Principle, Pluralism and Deliberation: Science and Ethics
(Responsible Research and Innovation Set – Volume 4)

Printed and bound by CPI Group (UK) Ltd, Croydon, CR0 4YY

16/09/2026

14963077-0002